基于转录组学技术的微塑料对有机污染物的斑马鱼毒性影响研究

侯 静 著

中国矿业大学出版社
·徐州·

内 容 提 要

本书主要介绍了水环境中微塑料与有机污染物(四溴双酚 A、全氟辛酸)联合作用对水生模式生物斑马鱼的毒性效应与作用机制,利用转录组学测序技术从分子水平揭示上述污染物的毒性机理。书中全面总结了作者多年来在相关领域的研究成果,特别是以模式生物斑马鱼为研究对象,通过生理生化与组织病理学分析,明确微塑料与新型污染物联合作用对斑马鱼不同组织的毒性差异;利用转录组学分析技术,比较微塑料与新型污染物联合作用所引起的基因表达调控模式变化,明确代谢通路、共表达网络与毒性差异之间的关系,解析微塑料对水环境中新型污染物在斑马鱼体内毒性效应影响的作用机制。

本书适合环境科学、化学、生物学、生态学、毒理学等相关领域的研究工作者阅读。

图书在版编目(ＣＩＰ)数据

基于转录组学技术的微塑料对有机污染物的斑马鱼毒性影响研究 / 侯静著. — 徐州：中国矿业大学出版社，2024.11. — ISBN 978-7-5646-6203-5

Ⅰ.X705;Q959.460.3

中国国家版本馆 CIP 数据核字第 2024ES9326 号

书　　名	基于转录组学技术的微塑料对有机污染物的斑马鱼毒性影响研究
著　　者	侯　静
责任编辑	何晓明　徐　玮
出版发行	中国矿业大学出版社有限责任公司
	(江苏省徐州市解放南路　邮编221008)
营销热线	(0516)83885370　83884103
出版服务	(0516)83995789　83884920
网　　址	http://www.cumtp.com　**E-mail**:cumtpvip@cumtp.com
印　　刷	苏州市古得堡数码印刷有限公司
开　　本	787 mm×1092 mm　1/16　**印张** 9.75　**字数** 191 千字
版次印次	2024 年 11 月第 1 版　2024 年 11 月第 1 次印刷
定　　价	42.00 元

(图书出现印装质量问题,本社负责调换)

前　言

自 1860 年第一种塑料诞生后的百余年内，塑料的需求量便因其优异的性能而迅速增加，然而其广泛的应用也衍生了诸多环境污染问题。微塑料便是由塑料垃圾的破碎或塑料微粒的不当处理而产生的，已被广泛证实会造成炎症反应、肝肠损伤及 DNA 损伤等，给人类带来的健康风险不容忽视。四溴双酚 A 作为典型的内分泌干扰物及致癌物，是一类被广泛添加于印刷线路板和电子元件中的卤系阻燃剂，2017 年已被世界卫生组织列入 2A 类致癌物清单。全氟辛酸作为四氟乙烯聚合及氟橡胶生产的分散剂，已在我国多处的地表水及地下水环境中被检出，因其致癌作用及内分泌干扰作用被列入我国 2023 年度重点管控的新污染物清单。

微塑料因其较高的比表面积和疏水性，可以作为其他有机污染物的载体，进而改变有机污染的毒理特性。现有研究表明，微塑料对不同类型有机污染物的毒性影响存在较大的差异。因此，有必要寻找适合技术评价微塑料对有机污染物的毒性影响。转录组学技术由于其精确性、高效性及可扩展性已经迅速成为污染物毒性影响评价的重要方法，其作用是鉴定特定生物相对于对照的差异表达基因，即确定对于选定基因，实验组样本的表达水平相对于对照组样本是否存在显著的差异。在微塑料对有机污染物的毒性影响这一课题中，转录组测序可以鉴定在特定污染物诱导下受试生物差异表达的基因，进而将这种基因的差异表达模式与生物细胞、组织和生物体的表型反应相结合，从分子层面预测其毒性机制，这成为解决这一问题的有效方法。

基于上述内容，本书通过转录组测序技术与其他毒理学研究手段相结合的方法，分别开展了微塑料对四溴双酚 A 及微塑料对全氟辛酸的斑马鱼毒性影响研究。相关结果表明，不同种类的微塑料对于两种污染物的不同毒性具有不同的效应。本书的研究受到了国家自然科学基金"微塑料对典型疏水性有机污染物在斑马鱼体内毒性效应的影响机制"（编号：22276052）的支持。吴浩迪和张思怡参与了本书的撰写与校对，在此向他们表示感谢。本书研究成果有助于了解微塑料影响有机污染物水生生物毒性的作用方式，评估它们对水生生态系统乃至食物链更高营养级的潜在威胁和生态风险，同时也能为科学制定水体污染防治策略提供理论支撑。

当然，本书所述的相关研究涉及领域众多，书稿难免存在疏漏或不足，在此敬请广大读者批评指正。

著 者
2024 年 7 月

目 录

第1章 微塑料对四溴双酚A的毒性影响综述 ············· 1
  1.1 选题背景及研究意义 ························· 1
  1.2 国内外研究现状 ···························· 3
  1.3 基于斑马鱼模型的毒性效应评估现状 ··············· 8
  1.4 研究目的与内容 ···························· 9
  参考文献 ································· 11

第2章 聚乙烯微塑料对四溴双酚A斑马鱼毒性效应影响研究 ······ 19
  2.1 引言 ··································· 19
  2.2 实验材料与方法 ··························· 19
  2.3 TBBPA的急性毒性 ························· 25
  2.4 PE-MPs的粒径与形貌特征 ··················· 26
  2.5 暴露溶液中TBBPA浓度 ····················· 27
  2.6 TBBPA与PE-MPs对斑马鱼生化指标的影响 ········ 28
  2.7 TBBPA与PE-MPs对斑马鱼组织的影响 ·········· 30
  2.8 本章小结 ······························· 31
  参考文献 ································· 32

第3章 聚乙烯微塑料对四溴双酚A斑马鱼毒性影响机制 ········ 34
  3.1 引言 ··································· 34
  3.2 实验材料与方法 ··························· 34
  3.3 转录组测序实验质量控制与结果 ················· 39
  3.4 TBBPA对斑马鱼毒性影响的潜在机理 ············· 42

3.5　PE-MPs 对 TBBPA 斑马鱼毒性影响的潜在机理 ·················· 52
　　3.6　转录组测序结果验证 ·· 69
　　3.7　TBBPA 对斑马鱼内分泌干扰效应验证 ·························· 70
　　3.8　本章小结 ·· 73
　　参考文献 ·· 74

第 4 章　微塑料对全氟辛酸的毒性影响综述 ······························ 78
　　4.1　选题背景 ·· 78
　　4.2　国内外研究现状 ··· 80
　　4.3　研究内容及意义 ··· 83
　　参考文献 ·· 85

第 5 章　聚苯乙烯微塑料对全氟辛酸斑马鱼毒性效应影响研究 ······ 95
　　5.1　引言 ·· 95
　　5.2　实验材料与方法 ··· 95
　　5.3　PFOA 与 PS-MPs 的急性毒性 ······································ 99
　　5.4　PFOA 含量测定及富集分析 ·· 102
　　5.5　PFOA 与 PS-MPs 对斑马鱼组织的影响 ·························· 105
　　5.6　本章小结 ··· 108
　　参考文献 ··· 108

第 6 章　聚苯乙烯微塑料对全氟辛酸斑马鱼毒性影响机制 ··········· 111
　　6.1　引言 ··· 111
　　6.2　实验材料与方法 ·· 111
　　6.3　PFOA 对斑马鱼毒性影响的潜在机理 ···························· 113
　　6.4　PS-MPs 与 PFOA 对斑马鱼的毒性影响的潜在机理 ··········· 119
　　6.5　不同粒径 PS-MPs 对 PFOA 斑马鱼的毒性影响的差异 ······· 134
　　6.6　本章小结 ··· 147
　　参考文献 ··· 148

第1章 微塑料对四溴双酚A的毒性影响综述

1.1 选题背景及研究意义

自1860年人类发明了第一种塑料以来,这种材料的需求量便因其优异的性能而飞速增加。从1950年到2022年,全世界塑料总产量从200万t增长到了4.003亿t[1-2]。预计20年后,塑料产量将再翻一倍[3]。在推动社会进步的同时,塑料的不当处理带来了严重的环境问题。据估计,塑料废物正在以2.74亿~2.75亿t/a的速度产生,而到2050年将共计约有120亿t的塑料垃圾进入环境中[4-5]。进入环境中的塑料垃圾除了污染环境外,还会在风力、水力、洋流、紫外辐射和微生物等因素的作用下破碎,生成更微小、更易于扩散的微塑料(MPs)[6]。MPs是粒径小于5 mm的塑料[7],除了塑料垃圾的降解破碎之外,以特殊用途制造的塑料微粒也是MPs污染的来源,如化妆品及工业用塑料微珠等[8]。近年来,随着塑料的产量及废弃量不断上升,加之新冠疫情的流行造成了全球口罩使用量的剧增,MPs污染扩散风险与日俱增。据估计,疫情期间全球每月最高可消费1 290亿个一次性口罩,而用于制造这些口罩的塑料占了全球塑料总产量的1.05%,约370万t[9]。作为已被查明的MPs产生源头,这些口罩的不当废弃导致了MPs污染更为广泛的传播。目前已在水体、土壤和大气等环境中被检出,而水环境更是MPs污染的重灾区[10-11]。水生浮游动物、鱼类和甲壳类动物中同样发现了的MPs的累积[12-14]。摄入MPs会给水生生物带来诸多不利影响,包括诱导机体炎症反应或介导氧化应激,引起DNA损伤及凋亡级联反应,乃至肝肠损伤、体重减轻甚至死亡[15]。除此之外,由于其疏水性及较高的比表面积,MPs还能作为其他污染物的载体,进而改变其毒理特性[16]。同时,考虑到MPs可以沿食物链富集,其带来的人体健康风险也不容忽视[17]。因此,MPs及其相关污染物的毒理学研究应得到人们的重视。

四溴双酚A(TBBPA)(图1-1)是塑料、纺织品和电子产品中最常用的卤系阻燃剂,其销售额约占传统阻燃剂市场的60%,并仍在不断上升。预计到2025年其营业额将上升至35亿美元[18-19]。研究表明,亚洲地区占全球TBBPA需求量的80%,而我国是印刷线路板和电子元件的重要生产国,具有可观的TBBPA需求[20]。我国除了进口TBBPA之外,在天津、山东及江苏等地均有大

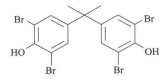

图 1-1　TBBPA 的分子结构

量的TBBPA生产工厂[21],TBBPA的合成及相关产品的制造因此成为我国TBBPA污染的重要来源。TBBPA可以通过多种途径进入环境,包括工业合成过程中的排放、含TBBPA产品在使用过程中的释放、含TBBPA相关产品的回收过程以及垃圾处理场中TBBPA的浸出。在我国的空气、灰尘、土壤、水体和沉积物中均检测到了TBBPA的存在[22-25],其中水体是TBBPA污染的重灾区。而在人体中也有TBBPA被检出[26]。现有研究表明,由于TBBPA与甲状腺激素T_4的相似性,所以其展现出内分泌干扰作用,能影响正常的细胞功能[27],同时TBBPA还被证明是雌激素类似物,能够影响生物的生殖功能[28],除此之外,TBBPA还能造成神经毒性及氧化应激等方面的毒性影响[20]。基于TBBPA对人体健康的威胁,目前世界上许多国家已经将TBBPA列为风险化学品并实施管理,美国环境保护署(EPA)已于2014年将TBBPA添加到有毒物质控制法(TSCA)工作计划清单中,丹麦和挪威也于2007年和2009年先后将TBBPA列为"不受欢迎"的物质[20]。考虑到我国法律尚未对TBBPA进行关注,对TBBPA的毒性进行评估有助于增进公众对这一化学品的认识,具有重要的现实意义。

由于TBBPA在塑料产品中的广泛应用,所以进入环境中的TBBPA通常会与MPs形成复杂的共污染体系,而基于复杂的相互作用,TBBPA在MPs存在与不存在的情况下会表现出不同的毒性效应。目前大多数研究表明,MPs会加重TBBPA的毒性效应。Zhang等[29]的研究表明,MPs会降低TBBPA商用文蛤解毒过程中丙酮酸激酶等酶的活性,导致解毒过程中能量供应不足,从而中断解毒过程,最终导致TBBPA的生物富集。Yu等[30]对TBBPA和MPs对斑马鱼的亚慢性暴露研究也表明,联合暴露增加了斑马鱼体内抗氧化酶的活性,诱导了比MPs或TBBPA单独暴露更高的抗氧化应激反应。然而,也有一些研究表明MPs可以降低一些有机污染物的毒性作用。例如,MPs可暂时提高芘在斑马鱼体内的代谢率,并在一定浓度下显著延缓芘对虾虎鱼的致死效应[31]。而MPs的浓度可能是造成这种差异的重要因素。考虑到现有的研究多针对单一浓度MPs对TBBPA毒性的影响,可能无法全面揭示MPs对TBBPA毒性的影

响趋势。因此,本研究通过设置 MPs 浓度梯度研究 MPs 对 TBBPA 毒理效应的影响,揭示 MPs 与 TBBPA 在水生生物体内的毒性作用方式,有助于评估它们对水生生态系统乃至食物链更高营养级的潜在威胁和生态风险,进而能为水体污染防治策略的制定提供理论支撑,具有重要的科学与现实意义。

1.2 国内外研究现状

1.2.1 水环境中的 TBBPA 及其影响

作为最常用的阻燃剂,TBBPA 的大量生产导致其已经在世界各地水环境中被检出。Labadie 等[32]发现,法国 Predecelle 河水与沉积物中的 TBBPA 丰度分别为 0.050~0.064 ng/g 和 0.065~0.130 ng/g;Guerra 等[33]发现,在西班牙的 Ebro 河流域沉积物中 TBBPA 丰度为 0.6~2.7 ng/g;Zhang 等[34]的研究则表明新加坡海洋沉积物中的 TBBPA 丰度为 0.2 ng/g。中国作为电子产品的主要生产国,具有庞大的 TBBPA 消费量,这就导致了我国水环境中的 TBBPA 污染情况较为严重。Xu 等[35]的研究表明,太湖水体沉积物中的 TBBPA 丰度为 0.06~2.12 ng/g;Xiong 等[36]的报告指出,中国广东某地地表水的 TBBPA 丰度最高达到了 920 ng/L;而 He 等[37]的研究表明,广东东江集水区水体 TBBPA 丰度平均为 1.75 ng/L。TBBPA 进入水环境的主要途径包括三种:① TBBPA 可能在生产过程中进入水环境,这导致了 TBBPA 生产地周围的水环境污染。山东寿光因其丰富的溴资源成为中国最大的 TBBPA 制造中心,其周边环境中较高的 TBBPA 丰度证实了 TBBPA 在生产过程的泄漏会造成环境污染[38]。② 电子废物的处理不当也是 TBBPA 的污染途径。TBBPA 被大量用于电路板的制造,而电子产品报废后的不当处理,如不正规的切割、磨碎、融化及露天燃烧等过程会导致大量含 TBBPA 烟雾、渗滤液、灰烬、颗粒物和废水的产生,进而对周边水环境产生污染[39]。③ 污水处理厂的废水也会导致 TBBPA 进入水环境。污水处理厂的生活及工业废水中 TBBPA 的丰度较高,其废水及污泥是水环境 TBBPA 的重要来源[23]。

研究表明,TBBPA 会对水生生物造成神经毒性、发育、内分泌干扰、免疫应激和生殖障碍等影响。例如,Yu 等[40]对斑马鱼的 42 d 的暴露实验表明,TBBPA 可致子代幼鱼甲状腺激素 T_3 和多巴胺指标异常,并下调多巴胺相关基因 *Drd2b* 和 *Drd3* 的表达水平,这一结果表明 TBBPA 对斑马鱼产生了神经毒性和内分泌毒性。Zhu 等[41]的 144 h 暴露实验结果则表明,TBBPA 在造成斑马鱼体内功能异常、干扰甲状腺激素正常作用的同时,显著下调了 L-tubulin 等神经相关基因的表达,这也验证了 TBBPA 的神经毒性和内分泌毒性。在发育毒性方面,Yang 等[42]的研究表明,TBBPA 暴露显著提高了斑马鱼胚胎及幼鱼的死亡率,导致了斑马鱼过量活性氧的产生及血液流动障碍,并显著增加了斑马鱼的畸形率。Usenko 等[43]以斑马鱼胚胎为研究对象的相似研究也表明,TBBPA 暴露可通过影响下丘脑-垂体-甲状腺(HPT)轴基因表达进而使斑马鱼胚胎运动速度下降,且畸形率和死亡率明显提高。免疫毒性方面,Gong 等[44]的研究表明,TBBPA 暴露抑制了鲤鱼细胞外诱捕网(NETs)的产生,同时激活了鲤鱼的线粒体细胞凋亡途径与坏死性凋亡途径,表明其影响了鲤鱼的正常免疫功能。Zhang 等[45]发现 TBBPA 暴露显著造成了黑斑蛙的睾丸组织损伤,增加了其精子畸形率,同时显著上调了其体内的性激素水平,表明了 TBBPA 存在生殖毒性。除此之外,已有证据表明水环境中的 TBBPA 可以通过食物链影响人类的健康,这使得 TBBPA 污染受到了人们的重视[38](图 1-2)。

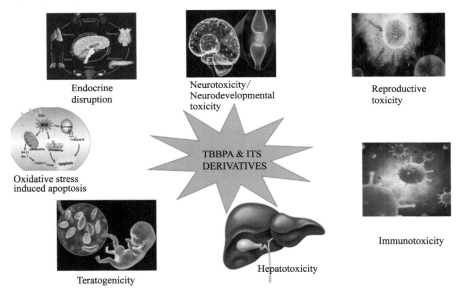

图 1-2　TBBPA 污染对人类的健康影响[38]

1.2.2　水环境中的 MPs 及其影响

水环境是 MPs 污染的重要聚集地。Carpenter 等[46]在 1972 年于《科学》杂志上报道了大量直径为 0.1～0.2 mm 的聚苯乙烯(PS)颗粒在英格兰沿海被发现,这也是最早的关于微型塑料污染的研究。而随着近年来塑料产量的不断增加,MPs

污染的分布也日益广泛。我国是 MPs 污染的重灾区,目前已经在我国的海洋、江河和湖泊等水环境中发现了 MPs 污染[47]。例如,南海西沙附近海域沉积物中 MPs 的丰度为 5～20 个/kg[48];Huang 等[49]的研究表明,我国长江口的水产养殖塘中水和沉积物中 MPs 丰度分别约为 36.2 个/L 和 271.65 个/kg,污染负荷指数均达到了风险Ⅰ级;而 Xiong 等[50]的研究表明,洪湖水域的 MPs 丰度达到了 117～533 个/m³。而水环境中的 MPs 会被水生动物摄入和累积,这在针对 MPs 污染的研究中得到了证明。Mohsen 等[51]在渤海和黄海的海参中发现了 MPs,其丰度为 1.56～24.2 个/条;对上海崇明岛的养殖鳗鱼、泥鳅和小龙虾的研究表明,这些养殖产品中的 MPs 平均丰度约为(1.7±0.5) 个/个体[52];而 Zhu 等[53]在烟台的牡蛎养殖场中也发现了 MPs,丰度为 4.53 个/单位组织重量,这表明 MPs 污染已经影响到了水生动物,而人类对这些水产品的消费有摄入 MPs 的风险,因此,环境中的 MPs 污染不能忽视(图 1-3)。

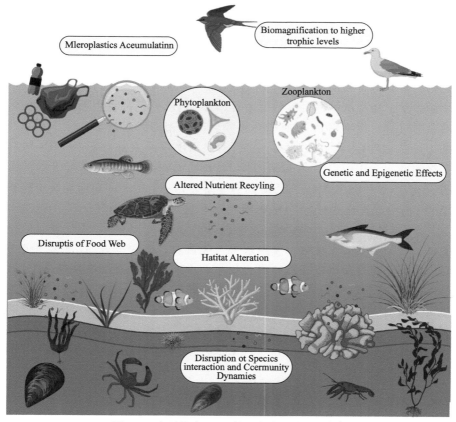

图 1-3 水环境中 MPs 的生态毒理学影响[54]

现有研究表明，MPs对水生生物存在神经、免疫和生殖等方面的毒理影响。MPs会影响乙酰胆碱酯酶（AchE）的正常功能，进而影响斑马鱼的神经系统[55]，主要表现为神经元坏死和细胞质空泡化，进而诱导斑马鱼进食减少与呼吸异常[56]。在免疫层面上，研究表明即使粒径达 $180\sim600~\mu m$ 的MPs颗粒也能进入欧洲凤尾鱼的肝脏组织[57]。MPs在肝脏内的诱导氧化应激是其产生免疫毒性的主要途径。研究表明，MPs会抑制青鳉鱼体内抗氧化酶的活性，诱发肝脏氧化应激甚至导致炎症，进而诱导糖脂代谢和能量代谢紊乱，最终影响其免疫功能[58]。此外，MPs也会影响斑马鱼的生殖系统。Sarasamma等[59]的研究表明，斑马鱼暴露在MPs中一个月后在鱼的生殖腺中有MPs的分布，并影响了其生殖功能。考虑到食物链对污染物的富集效应，人类具有较高的MPs摄入风险，因此MPs的毒理效应应当得到进一步评估。

1.2.3 MPs对其他污染物的毒理影响

MPs可以通过分配作用、表面吸附、交互作用和其他微观机制等吸附机理吸附有机污染物，其中表面吸附是最主要的作用机制。大多数MPs颗粒对有机污染物的吸附平衡是MPs聚合物表面吸附和内部无定形区域分配的综合效应。当有机污染物浓度较高时，MPs表面结合位点会很快发生饱和。在此之后，无论从动力学上还是从平衡时的浓度来看，MPs内部分配都将主导其对有机污染物的吸附过程。影响MPs与有机污染物吸附作用的因素较多。除MPs粒径、温度、pH值之外，水流速度、化学性质、老化程度等因素都会影响MPs与有机污染物间的相互作用[60]。针对聚丙烯MPs对 $3,3',4,4'$-四氯联苯吸附的研究表明，在动态水环境中MPs对污染物的吸附达到平衡的时间要明显快于静态水环境[61]，说明水流速度对MPs吸附能力具有显著影响。Hüffer等[62]针对多种微MPs对非极性有机化合物吸附效果进行了分析，结果显示MPs表面的化学性质以及污染物的疏水性对于吸附效果至关重要。研究也表明，自然环境中基于光氧化的老化过程可以通过在聚合物结构中引入极性基团来降低MPs的疏水性，进而影响其对污染物的分配系数，最终降低了MPs的吸附能力[63]。关于老化聚苯乙烯MPs对污染物吸附能力的分析结果也证明了这一点[64]，说明老化程度对MPs吸附能力也至关重要。此外，MPs与水生环境相互作用后，往往会在其表面形成一层生物膜，这种生物膜会降解多环芳烃和多氯联苯，干扰MPs对其的吸附过程[65]。

1.2.4 MPs与有机污染物的联合毒性

由于这些复杂的相互作用，目前研究者普遍认为MPs可以作为有机污染物

第 1 章 微塑料对四溴双酚 A 的毒性影响综述

积累和转移到水生环境中的载体,从而增加有机污染物在生物体内的累积风险,引起水生生物在分子、细胞、组织以及行为等多个方面的变化。在分子层面,Lin 等[66]的研究表明,聚苯乙烯 MPs 增强了多环芳烃在大型蚤肠道中脂质的传质效率,提高了多环芳烃的蓄积水平,使其对大型蚤的毒性显著增强。而 Zhang 等[67]发现聚苯乙烯 MPs 的存在显著增强了红罗非鱼组织中罗红霉素的生物积累,导致其肝脏的 EROD 和 BFCOD 两种酶的活性改变,进而导致红罗非鱼的代谢紊乱,产生氧化应激反应。Avio 等[68]研究发现,载有多环芳烃的聚乙烯和聚苯乙烯 MPs 会进入贻贝的消化组织和鳃,造成免疫反应、神经毒性和遗传毒性等不良影响。针对青鳉鱼的研究则表明,被多氯联苯、多环芳烃和多溴二苯醚污染的聚乙烯 MPs 可引起这些污染物在鱼体内的蓄积,引起糖原消耗、脂肪消耗和细胞坏死等一系列症状[69]。在行为方面,MPs 能够促进壬基酚、菲及三氯生等常见聚合物添加剂向蠕虫的转移效应,最终影响其行为与存活率[70]。而最近的研究表明,MPs 对有机污染物的毒性效应可能由于其浓度的不同而不同。例如,一项研究表明,当浓度低于 1 mg/L 时,MPs 浓度的升高降低了 PCB-18 对大型水蚤的致死效应,而当浓度高于 1 mg/L 时,MPs 浓度的升高加重了 PCB-18 的致死效应[71]。由此可见,MPs 对有机污染物的毒性效应的影响并非只有简单的加重作用,需要进一步的研究以加深人们对这一课题的认识。而 TBBPA 作为塑料添加剂,常与 MPs 共存进而形成复合污染[72],这造成了更复杂的污染效应,因此研究 MPs 对 TBBPA 的毒理影响是具有现实意义的(图 1-4)。

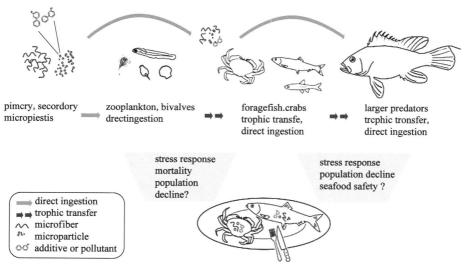

图 1-4　MPs 和有机污染物联合毒性作用对水生生物的影响[73]

1.3 基于斑马鱼模型的毒性效应评估现状

1.3.1 基于斑马鱼模型的毒理学评价方法

评估 MPs 对 TBBPA 的毒性效应影响需要选择合适的受试生物。考虑到本研究的背景为评估水环境中 MPs 对 TBBPA 毒性效应的影响,斑马鱼(*Danio rerio*)被选为此次研究的受试物种。1965 年,斑马鱼首次被用于测试硫酸锌的毒性[74],并在随后数十年里被广泛地应用于化学品的毒理研究中。2000—2002 年,研究人员开发了更适用于毒理学研究的转基因斑马鱼[75-76]。随后在长期的研究下,斑马鱼成为毒理学研究中的模式生物。斑马鱼在毒理学研究中具有多种不可替代的优势。首先,斑马鱼具有较短的繁殖周期,以体外受精的方式繁殖,这意味着斑马鱼在跨代毒性研究方面有着明显的优势;其次,斑马鱼基因组序列已于 2013 年公布,结果表明约 70% 的基因与人类同源[77],这意味着斑马鱼模型在基因层面可以有效反映人类对有毒物质的一些反应;再次,斑马鱼作为一种应用广泛的模式生物,具有成熟的研究方法以及流程体系。基于上述理由,本研究采用斑马鱼作为受试生物,以研究 MPs 对 TBBPA 的毒理效应的影响。

1.3.2 基于转录组测序的毒理学评价方法

转录组测序(RNA-Seq)是近十年来新兴的基于测序的基因表达测定技术。其作用是鉴定特定生物相对于对照的差异表达基因,即确定对于选定基因,实验组样本的表达水平相对于对照组样本是否存在显著的差异。相较于同为基因表达水平测定技术的微阵列技术,RNA-Seq 具有多方面的优势。例如,在可以高度精确地同时测量大量选定基因的表达量的同时,RNA-Seq 还可以研究选择性剪接、新转录本表达、等位基因特异性表达、基因融合事件和遗传变异等事件[78],因此目前 RNA-Seq 技术已经迅速成为转录组研究的首选方法。

RNA-Seq 的主要流程如下:首先从样本中提取总 RNA,其次通过逆转录将其转化为互补 DNA(cDNA),经处理后使用高通量测序技术获得原始 reads 数据集,随后将 reads 与已知转录组数据库中带注释的参考序列对齐以获得样本的基因组或转录组的表达水平。在毒理学研究中,RNA-Seq 可以鉴定在特定化学品诱导下受试生物差异表达的基因,进而将这种基因的差异表达模式与生物细胞、组织和生物体的表型反应相结合,进而从基因层面揭示化学品的致毒机理,因此其已经被广泛用于毒理学研究领域。

1.3.3 基于计算化学的毒理学评价方法

随着计算机技术的迅猛发展,计算机模拟方法已被广泛用于预测生物修复现象、清理污染物、评估毒性以及预测对有毒化合物降解可能性等领域。分子对接及分子动力学模拟是计算机模拟方法中的重要技术。分子对接是通过计算机模拟来寻找目标分子与目标蛋白最优结合构象的方法,该方法首先通过输入蛋白质及化学分子的结构评估结合位点,随后通过特定的评分函数选择结合模式进行对接,并在对接中产生的一系列可能的复合体中寻找可用于预测结合亲和力的最优结合构象[79]。分子动力学模拟是一种基于牛顿运动定律和统计力学原理的计算技术,主要用于模拟原子和分子随时间推移的行为和相互作用。在毒理学研究中,分子动力学模拟可以通过对生物体内环境的模拟来呈现选定化学物质与生物体内蛋白质结合的微观动态行为。构建分子动力学模拟所需的必要元素包括力场、系综和水模型[80]。力场是用于描述蛋白质分子和配体之间力的类型和大小的模型;系综用于确保模拟的温度和压力条件;水模型则用于蛋白质分子和配体的溶剂化,以表征它们在水中的结构状态。由于 TBBPA 与多种生物激素具有结构相似性,是典型的内分泌干扰物质,因此分子对接与分子动力学模拟的方法可以有效地从分子层面揭示 TBBPA 的致毒机理,有助于对 TBBPA 毒性效应的更全面的评估。

1.4 研究目的与内容

1.4.1 研究目的

① 解析 TBBPA 对斑马鱼的毒理效应。
② 阐述 MPs 对 TBBPA 毒性的影响。
③ 揭示 MPs 影响 TBBPA 毒性的机理。

1.4.2 研究内容

研究技术路线如图 1-5 所示。

本研究以斑马鱼为研究对象,选择聚合物阻燃剂 TBBPA 为目标污染物,评估其对斑马鱼的毒性效应;引入聚乙烯微塑料(PE-MPs),探究其存在对 TBBPA 毒性效应的影响。选择 PE-MPs 的原因是聚乙烯塑料在使用广泛的同时抗氧化能力较差,在环境中易于老化和粉碎形成 MPs,是水环境中分布最为

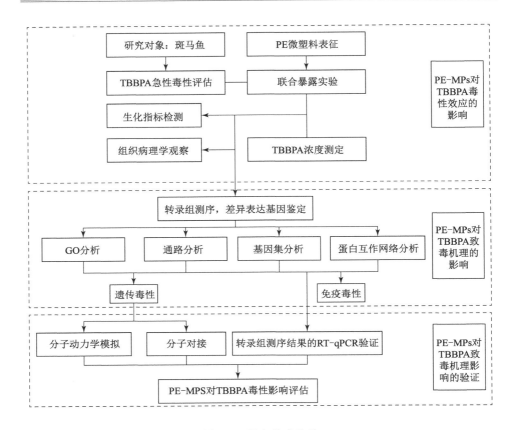

图 1-5 研究技术路线

广泛的 MPs。研究表明,在底栖鱼类和中上层鱼类的胃肠道中 PE-MPs 的富集量最大,远远超过任何其他类型的 MPs,因此在环境中 PE-MPs 有更高的可能影响 TBBPA 的毒理效应。本研究的主要内容如下:

① PE-MPs 对 TBBPA 的毒理效应的影响评估。通过 96 h 急性毒性实验评估 TBBPA 对斑马鱼的毒性影响,并确定后续联合暴露实验中 TBBPA 的暴露浓度,此外通过表征确定 PE-MPs 的表面特性,通过生化指标检测确定 PE-MPs 存在/不存在条件下 TBBPA 对斑马鱼 SOD、CAT 及 MDA 相关指标的影响,并通过组织病理学观察定位 PE-MPs 存在/不存在条件下 TBBPA 影响的斑马鱼组织。

② PE-MPs 对 TBBPA 毒性效应的影响机理评估。通过 RNA-Seq 鉴定 TBBPA 单独暴露体系及 TBBPA 与 PE-MPs 共暴露体系诱导的差异表达基因,采用生物信息学方法,通过 GO 富集分析、通路富集分析、基因集分析及蛋白互作网络分析等手段从基因层面揭示 TBBPA 的致毒机理及 PE-MPs 存在情况下

③ PE-MPs 对 TBBPA 毒性效应的影响机理验证。采用 RT-qPCR 验证 RNA-Seq 结果,并在生物信息学分析结果的基础上,通过分子对接与分子动力学模拟研究 TBBPA 对甲状腺激素受体 β(THRβ)和 17β-羟基类固醇脱氢酶 1(HSD17β1)两种蛋白的干扰作用,以全面揭示相关效应的作用机制。

参考文献

[1] RAFEY A,SIDDIQUI F Z.A review of plastic waste management in India-challenges and opportunities[J].International journal of environmental analytical chemistry,2023,103(16):3971-3987.

[2] NAYANATHARA T P,RATNAYAKE A S.The world of plastic waste:a review[J].Cleaner materials,2024,11:100220.

[3] LEBRETON L,ANDRADY A.Future scenarios of global plastic waste generation and disposal[J].Palgrave communications,2019,5(1):6.

[4] GEYER R,JAMBECK J R,LAW K L.Production,use,and fate of all plastics ever made[J].Science advances,2017,3(7):1700782.

[5] JAMBECK J R,GEYER R,WILCOX C,et al.Plastic waste inputs from land into the ocean[J].Science,2015,347(6223):768-771.

[6] AUTA H S,EMENIKE C U,FAUZIAH S H.Distribution and importance of microplastics in the marine environment:a review of the sources,fate,effects,and potential solutions[J].Environment international,2017,102:165-176.

[7] THOMPSON R C.Microplastics in the marine environment:sources,consequences and solutions[M]// Marine Anthropogenic Litter.Cham:Springer,2015:185-200.

[8] DABROWSKA A,MIELANCZUK M,SYCZEWSKI M.The Raman spectroscopy and SEM/EDS investigation of the primary sources of microplastics from cosmetics available in Poland[J].Chemosphere,2022,308(3):136407.

[9] ZHAO X,GAO P P,ZHAO Z Q,et al.Microplastics release from face masks:characteristics,influential factors,and potential risks[J].Science of the total environment,2024,921:171090.

[10] BOTTERELL Z L R,BEAUMONT N,DORRINGTON T,et al.Bioavailability and effects of microplastics on marine zooplankton:a review[J].Environmental pollution,2019,245:98-110.

[11] BARNES D K A, GALGANI F, THOMPSON R C, et al. Accumulation and fragmentation of plastic debris in global environments[J]. Philosophical transactions of the royal society of London series B, biological sciences, 2009, 364(1526):1985-1998.

[12] DESFORGES J P W, GALBRAITH M, ROSS P S. Ingestion of microplastics by zooplankton in the northeast Pacific Ocean[J]. Archives of environmental contamination and toxicology, 2015, 69(3):320-330.

[13] REED S, CLARK M, THOMPSON R, et al. Microplastics in marine sediments near Rothera Research Station, Antarctica[J]. Marine pollution bulletin, 2018, 133:460-463.

[14] KUMAR R, VERMA A, RAKIB M R J, et al. Adsorptive behavior of micro(nano)plastics through biochar: co-existence, consequences, and challenges in contaminated ecosystems[J]. Science of the total environment, 2023, 856(1):159097.

[15] RAFA N, AHMED B, ZOHORA F, et al. Microplastics as carriers of toxic pollutants: source, transport, and toxicological effects[J]. Environmental pollution, 2024, 343:123190.

[16] DING T D, WEI L Y, HOU Z M, et al. Microplastics altered contaminant behavior and toxicity in natural waters[J]. Journal of hazardous materials, 2022, 425:127908.

[17] WINIARSKA E, JUTEL M, ZEMELKA-WIACEK M. The potential impact of nano-and microplastics on human health: understanding human health risks[J]. Environmental research, 2024, 251(2):118535.

[18] LAW R J, ALLCHIN C R, DE BOER J, et al. Levels and trends of brominated flame retardants in the European environment[J]. Chemosphere, 2006, 64(2):187-208.

[19] WANG Z, PENG X, JIA X. Removal of tetrabromobisphenol A(TBBPA) by synergistic integration of anaerobic microbes and Fe-Ni bimetallic particles[J]. Acta entiae circumstantiae, 2018, 38(7):2607-2614.

[20] LIU K, LI J, YAN S J, et al. A review of status of tetrabromobisphenol A(TBBPA) in China[J]. Chemosphere, 2016, 148:8-20.

[21] MA J, QIU X H, ZHANG J L, et al. State of polybrominated diphenyl ethers in China: an overview[J]. Chemosphere, 2012, 88(7):769-778.

[22] FENG A H, CHEN S J, CHEN M Y, et al. Hexabromocyclododecane

(HBCD) and tetrabromobisphenol A(TBBPA) in riverine and estuarine sediments of the Pearl River Delta in Southern China, with emphasis on spatial variability in diastereoisomer-and enantiomer-specific distribution of HBCD[J].Marinepollution bulletin,2012,64(5):919-925.

[23] ZHOU X Y,GUO J,ZHANG W,et al.Tetrabromobisphenol A contamination and emission in printed circuit board production and implications for human exposure[J].Journal of hazardous materials,2014,273:27-35.

[24] WU Y Y,LI Y Y,KANG D,et al.Tetrabromobisphenol A and heavy metal exposure via dust ingestion in an e-waste recycling region in Southeast China[J].Science of the total environment,2016,541:356-364.

[25] ZHU Z C,CHEN S J,ZHENG J,et al.Occurrence of brominated flame retardants (BFRs),organochlorine pesticides (OCPs),and polychlorinated biphenyls (PCBs) in agricultural soils in a BFR-manufacturing region of North China[J].Science of the total environment,2014,481:47-54.

[26] SHI Z X,JIAO Y,HU Y,et al.Levels of tetrabromobisphenol A,hexabromocyclododecanes and polybrominated diphenyl ethers in human milkfrom the general population in Beijing,China[J].Science of the total environment,2013, 452/453:10-18.

[27] GUYOT R,CHATONNET F,GILLET B,et al.Toxicogenomic analysis of the ability of brominated flame retardants TBBPA and BDE-209 to disrupt thyroid hormone signaling in neural cells[J].Toxicology,2014,325: 125-132.

[28] KUIPER R V,VAN DEN BRANDHOF E J,LEONARDS P E G,et al. Toxicity of tetrabromobisphenol A (TBBPA) in zebrafish (*Danio rerio*) in a partial life-cycle test[J].Archivesof toxicology,2007,81(1):1-9.

[29] ZHANG W X,TANG Y,HAN Y,et al.Microplastics boost the accumulation of tetrabromobisphenol A in a commercial clam and elevate corresponding food safety risks[J].Chemosphere,2022,292:133499.

[30] YU Y J,MA R X,QU H,et al.Enhanced adsorption of tetrabromobisphenol A (TBBPA) on cosmetic-derived plastic microbeads and combined effects on zebrafish[J].Chemosphere,2020,248:126067.

[31] OLIVEIRA M,RIBEIRO A,HYLLAND K,et al.Single and combined effects of microplastics and pyrene on juveniles (0+ group) of the common goby Pomatoschistus microps (Teleostei, Gobiidae) [J]. Ecologi-

calindicators,2013,34:641-647.

[32] LABADIE P,TLILI K,ALLIOT F,et al.Development of analytical procedures for trace-level determination of polybrominated diphenyl ethers and tetrabromobisphenol A in river water and sediment[J]. Analytical andbioanalytical chemistry,2010,396(2):865-875.

[33] GUERRA P,ELJARRAT E,BARCELÓ D.Simultaneous determination of hexabromocyclododecane, tetrabromobisphenol A, and related compounds in sewage sludge and sediment samples from Ebro River Basin (Spain)[J].Analyticaland bioanalytical chemistry,2010,397(7):2817-2824.

[34] ZHANG H,BAYEN S,KELLY B C.Co-extraction and simultaneous determination of multi-class hydrophobic organic contaminants in marine sediments and biota using GC-EI-MS/MS and LC-ESI-MS/MS[J].Talanta,2015,143:7-18.

[35] XU J,ZHANG Y,GUO C S,et al.Levels and distribution of tetrabromobisphenol A and hexabromocyclododecane in Taihu Lake,China[J].Environmental toxicology and chemistry,2013,32(10):2249-2255.

[36] XIONG J K,AN T C,ZHANG C S,et al.Pollution profiles and risk assessment of PBDEs and phenolic brominated flame retardants in water environments within a typical electronic waste dismantling region[J].Environmental geochemistry and health,2015,37(3):457-473.

[37] HE M J,LUO X J,YU L H,et al.Diasteroisomer and enantiomer-specific profiles of hexabromocyclododecane and tetrabromobisphenol A in an aquatic environment in a highly industrialized area,South China:vertical profile,phase partition,and bioaccumulation[J].Environmental pollution,2013,179:105-110.

[38] SUNDAY O E,HUANG B,MAO G H,et al.Review of the environmental occurrence,analytical techniques,degradation and toxicity of TBBPA and its derivatives[J].Environmental research,2022,206:112594.

[39] SEPÚLVEDA A,SCHLUEP M,RENAUD F G,et al.A review of the environmental fate and effects of hazardous substances released from electrical and electronic equipments during recycling:examples from China and India[J].Environmental impact assessment review,2010,30(1):28-41.

[40] YU Y J,HOU Y B,DANG Y,et al.Exposure of adult zebrafish (*Danio rerio*) to Tetrabromobisphenol A causes neurotoxicity in larval offspring, an adverse transgenerational effect[J].Journal of hazardous materials,

2021,414:125408.

[41] ZHU B R,ZHAO G,YANG L H,et al.Tetrabromobisphenol A caused neurodevelopmental toxicity via disrupting thyroid hormones in zebrafish larvae[J].Chemosphere,2018,197:353-361.

[42] YANG S W,WANG S R,SUN F C,et al.Protective effects of puerarin against tetrabromobisphenol a-induced apoptosis and cardiac developmental toxicity in zebrafish embryo-larvae[J].Environmental toxicology,2015,30(9):1014-1023.

[43] USENKO C Y,ABEL E L,HOPKINS A,et al.Evaluation of common use brominated flame retardant (BFR) toxicity using a zebrafish embryo model[J].Toxics,2016,4(3):21.

[44] GONG D Q,SUN K X,YIN K X,et al.Selenium mitigates the inhibitory effect of TBBPA on NETs release by regulating ROS/MAPK pathways-induced carp neutrophil apoptosis and necroptosis[J].Fish and shellfish immunology,2023,132:108501.

[45] ZHANG H J,LIU W L,CHEN B,et al.Differences in reproductive toxicity of TBBPA and TCBPA exposure in male Rana nigromaculata[J].Environmental pollution,2018,243(1):394-403.

[46] CARPENTER E J,ANDERSON S J,HARVEY G R,et al.Polystyrene spherules in coastal waters[J].Science,1972,178(4062):749-750.

[47] 周智勇,卢帅良,马涛,等.水环境微塑料污染现状及处置技术研究进展[J].净水技术,2024(7):10-21,156.

[48] HUANG L,LI Q P,LI H X,et al.Microplastic pollution and regulating factors in the surface sediment of the Xuande Atolls in the South China Sea[J].Marine pollution bulletin,2023,196:115562.

[49] HUANG J N,XU L,WEN B,et al.Characteristics and risks of microplastic contamination in aquaculture ponds near the Yangtze Estuary,China[J].Environmental pollution,2024,343:123288.

[50] XIONG X,LIU Q,CHEN X C,et al.Occurrence of microplastic in the water of different types of aquaculture ponds in an important lakeside freshwater aquaculture area of China[J].Chemosphere,2021,282:131126.

[51] MOHSEN M,WANG Q,ZHANG L B,et al.Microplastic ingestion by the farmed sea cucumber Apostichopus japonicus in China[J].Environmental pollution,2019,245:1071-1078.

[52] LV W W,ZHOU W Z,LU S B,et al.Microplastic pollution in rice-fish co-

culture system: a report of three farmland stations in Shanghai, China[J]. Science of the total environment, 2019, 652: 1209-1218.

[53] ZHU X T, QIANG L Y, SHI H H, et al. Bioaccumulation of microplastics and its in vivo interactions with trace metals in edible oysters[J]. Marine pollution bulletin, 2020, 154: 111079.

[54] THACHARODI A, MEENATCHI R, HASSAN S, et al. Microplastics in the environment: a critical overview on its fate, toxicity, implications, management, and bioremediation strategies[J]. Journal of environmental management, 2024, 349: 119433.

[55] TLILI S, JEMAI D, BRINIS S, et al. Microplastics mixture exposure at environmentally relevant conditions induce oxidative stress and neurotoxicity in the wedge clam *Donax trunculus*[J]. Chemosphere, 2020, 258: 127344.

[56] UMAMAHESWARI S, PRIYADARSHINEE S, BHATTACHARJEE M, et al. Exposure to polystyrene microplastics induced gene modulated biological responses in zebrafish (*Danio rerio*)[J]. Chemosphere, 2021, 281: 128592.

[57] COLLARD F, GILBERT B, COMPÈRE P, et al. Microplastics in livers of European anchovies (*engraulis encrasicolus*, L.)[J]. Environmental pollution, 2017, 229: 1000-1005.

[58] FENG S B, ZENG Y H, CAI Z H, et al. Polystyrene microplastics alter the intestinal microbiota function and the hepatic metabolism status in marine medaka (*Oryzias melastigma*)[J]. Science of the total environment, 2021, 759: 143558.

[59] SARASAMMA S, AUDIRA G, SIREGAR P, et al. Nanoplastics causeneurobehavioral impairments, reproductive and oxidative damages, and biomarker responses in zebrafish: throwing up alarms of wide spread health risk of exposure[J]. International journal of molecular sciences, 2020, 21(4): 1410.

[60] KARAPANAGIOTI H K, KLONTZA I. Testing phenanthrene distribution properties of virgin plastic pellets and plastic eroded pellets found on Lesvos Island beaches (Greece)[J]. Marine environment research, 2008, 65(4): 283-290.

[61] ZHAN Z W, WANG J D, PENG J P, et al. Sorption of 3,3′,4,4′-tetrachlorobiphenyl by microplastics: a case study of polypropylene[J]. Marine pollution bulletin, 2016, 110(1): 559-563.

[62] HÜFFER T, HOFMANN T. Sorption of non-polar organic compounds by micro-sized plastic particles in aqueous solution[J]. Environmental pollu-

tion,2016,214:194-201.

[63] ENDO S,TAKIZAWA R,OKUDA K,et al.Concentration of polychlorinated biphenyls (PCBs) in beached resin pellets:variability among individual particles and regional differences[J].Marine pollution bulletin,2005,50(10):1103-1114.

[64] HÜFFER T,WENIGER A K,HOFMANN T.Sorption of organic compounds by aged polystyrene microplastic particles[J].Environmental pollution,2018,236:218-225.

[65] JIN M,YU X B,YAO Z Y,et al.How biofilms affect the uptake and fate of hydrophobic organic compounds (HOCs) in microplastic:insights from anin situ study of Xiangshan Bay,China[J].Water research,2020,184:116118.

[66] LIN W,JIANG R F,XIAO X Y,et al.Joint effect of nanoplastics and humic acid on the uptake of PAHs for Daphnia Magna:a model study[J].Journal of hazardous materials,2020,391:122195.

[67] ZHANG S S,DING J N,RAZANAJATOVO R M,et al.Interactive effects of polystyrene microplastics and roxithromycin on bioaccumulation and biochemical status in the freshwater fish red tilapia (*Oreochromis niloticus*)[J].Science of the total environment,2019,648:1431-1439.

[68] AVIO C G,GORBI S,MILAN M,et al.Pollutants bioavailability and toxicological risk from microplastics to marine mussels[J].Environmental pollution,2015,198:211-222.

[69] ROCHMAN C M,KUROBE T,FLORES I,et al.Early warning signs of endocrine disruption in adult fish from the ingestion of polyethylene with and without sorbed chemical pollutants from the marine environment[J].Science of the total environment,2014,493:656-661.

[70] BROWNE M A,NIVEN S J,GALLOWAY T S,et al.Microplastic moves pollutants and additives to worms,reducing functions linked to health and biodiversity[J].Current biology,2013,23(23):2388-2392.

[71] LIN W,JIANG R F,XIONG Y X,et al.Quantification of the combined toxic effect of polychlorinated biphenyls and nano-sized polystyrene on Daphnia Magna[J].Journal of hazardous materials,2019,364:531-536.

[72] DO A T N,HA Y,KWON J H.Leaching of microplastic-associated additives in aquatic environments:a critical review[J].Environmental pollution,2022,305:119258.

[73] BAECHLER B R, STIENBARGER C D, HORN D A, et al. Microplastic occurrence and effects in commercially harvested North American finfish and shellfish: current knowledge and future directions[J]. Limnology andoceanography letters, 2020, 5(1): 113-136.

[74] SKIDMORE J F. Resistance to zinc sulphate of the zebrafish (*Brachydanio rerio* Hamilton-Buchanan) at different phases of its life history[J]. Theannals of applied biology, 1965, 56(1): 47-53.

[75] AMANUMA K, TONE S, SAITO H, et al. Mutational spectra of benzo[a]pyrene and MeIQx in rpsL transgenic zebrafish embryos[J]. Genetic toxicology and environmental mutagenesis, 2002, 513(1/2): 83-92.

[76] PATRICK BAUER M, WILLIAM GOETZ F. Isolation of gonadal mutations in adult zebrafish from a chemical mutagenesis Screen1[J]. Biology of reproduction, 2001, 64(2): 548-554.

[77] ZHAO W C, CHEN Y N, HU N, et al. The uses of zebrafish (*Danio rerio*) as an *in vivo* model for toxicological studies: a review based on bibliometrics[J]. Ecotoxicology and environmental safety, 2024, 272: 116023.

[78] CHOWDHURY H A, BHATTACHARYYA D K, KALITA J K. Differential expression analysis of RNA-seq reads: overview, taxonomy, and tools[J]. IEEE/ACM transactions on computational biology and bioinformatics, 2020, 17(2): 566-586.

[79] TU M C, ZHENG X, LIU P Y, et al. Typical organic pollutant-protein interactions studies through spectroscopy, molecular docking and crystallography: a review[J]. Science of the total environment, 2021, 763: 142959.

[80] SILVA M C, DE CASTRO A A, LOPES K L, et al. Combining computational tools and experimental studies towards endocrine disruptors mitigation: a review of biocatalytic and adsorptive processes[J]. Chemosphere, 2023, 344: 140302.

第 2 章　聚乙烯微塑料对四溴双酚 A 斑马鱼毒性效应影响研究

2.1　引言

目前 TBBPA 在水环境中的环境丰度不断增加,而 MPs 作为常与其共存的污染物,可能存在对 TBBPA 的毒性影响。本章以斑马鱼作为受试生物,以 TBBPA 与不同浓度的 PE-MPs 为暴露污染物,以急性毒性实验及联合暴露实验为基础,采用生化指标测定及组织病理学观察等手段对 TBBPA 的毒性及 PE-MPs 对其的影响进行初步探究。

2.2　实验材料与方法

2.2.1　受试生物

受试生物为 40 日龄健康野生型 AB 品系斑马鱼。斑马鱼购自上海费曦生物科技有限公司并在实验前利用曝气自来水驯养一周以上。驯养及实验过程遵照经济合作与发展组织(OECD)的鱼类急性毒性实验指南(OECD Test No.203)[1],并利用水质测定仪确定驯养期间的水质参数。驯养条件如下:水温(24±1) ℃,pH 值 7.4±0.6,电导率(317±10) μs/cm,溶解氧(6±1) ppm(10^{-6}),光照周期 12/12 h(光/暗)。驯养期间每天一次定时投喂斑马鱼丰年虾卵,喂食 30 min 后去除残余的食物和粪便,驯养期间斑马鱼死亡率小于 5%,暴露实验前 24 h 停止喂食。随机选取 20 条斑马鱼测定体长及体重,结果分别为(14.6±1.2) cm 和 (31.7±3.2) mg。

2.2.2 试剂与仪器

四溴双酚 A(TBBPA，$C_{15}H_{12}Br_4O_2$，CAS 79-94-7，纯度≥99%)和二甲基亚砜(DMSO，CAS 67-68-5，纯度≥99.9%)购自上海阿拉丁生化科技股份有限公司。100 μm PE-MPs(纯度≥99%)购自上海阳励机电科技有限公司。实验前将TBBPA溶于DMSO中制成浓度为 10 g/L 的 TBBPA 储备母液，并于 -18 ℃ 条件下储存待用。分析级乙醇(CAS 64-17-5，纯度≥99.7%)购自北京化工厂有限责任公司，冰醋酸(CAS64-19-7，纯度≥99.5%)和 4% 多聚甲醛固定液(CH_2O_n)购自上海麦克林生物科技股份有限公司。总蛋白(TP)测定试剂盒(No.A045-2)、超氧化物歧化酶(SOD)活性测定试剂盒(No.A001-1)、过氧化氢酶(CAT)活性测定试剂盒(No.A007-1)、丙二醛(MDA)含量测定试剂盒(No.A003-1)购自南京建成生物工程研究所。本章所用实验仪器见表 2-1。

表 2-1 实验仪器

仪器名称	仪器型号	生产厂家
超声波清洗器	KH-500DE	昆山禾创超声仪器有限公司
冰箱	BCD-452WDPF	青岛海尔股份有限公司
超低温冰箱	DW-86L100	青岛海尔生物医疗股份有限公司
超纯水系统	OKP-10	上海涞科仪器有限公司
电子天平	FA2204N	上海菁海仪器有限公司
高压灭菌器	MLS-378K-PC	Panasonic
水质检测仪	HI9829-04	Hanna Instruments
智能人工气候箱	RXZ-1500c-2	宁波江南仪器厂
高效液相色谱仪	1260 Infinity II	Agilent
环境扫描电子显微镜	Quattro C	ThermoFisher
涡旋振荡器	MIX-28	杭州米欧仪器有限公司
制冰机	FM40	北京长流科学仪器有限公司
恒温水浴锅	HW.SY21-KP6	北京市长风仪器仪表公司
高速冷冻离心机	5424R	Eppendorf
组织包埋机	HistoCore Arcadia H	Leica
数字切片扫描仪	MIDI	Pannoramic
酶标仪	1510	ThermoFisher

2.2.3 急性毒性实验

采用 96 h 静态暴露实验评估 TBBPA 对斑马鱼的致死毒性。将母液和脱氯自来水加入 500 mL 玻璃烧杯中配置暴露溶液,使每个烧杯中含有 400 mL 溶液。同时设立空白对照和 0.05%(v/v)的 DMSO 溶剂对照(各暴露溶液中的 DMSO 最高值)。根据预实验结果,浓度梯度设置为 0.5 mg/L、1 mg/L、2 mg/L、3 mg/L、4 mg/L 和 5 mg/L,每个实验浓度和对照重复 3 次。在每个烧杯中随机放入 10 条斑马鱼并在智能人工气候箱中进行 96 h 暴露,控制实验水温为 (24±1) ℃,光照周期为 12/12 h(光/暗),光照强度为 1 000 lux。实验期间不进行喂食或换水,每天测定水质参数并及时清理死亡的斑马鱼。在暴露实验结束后,利用 probit 模型(SPSS 27)确定不同时段 TBBPA 的半致死浓度(LC_{50})。

2.2.4 PE-MPs 表征

将干燥的 PE-MPs 样品固定于导电胶带上,在喷金处理后通过环境扫描显微镜(ESEM,Quattro C,ThermoFisher,USA)在 15 kV 的加速电压下进行观察,以表征 PE-MPs 的尺寸与表面形貌。

2.2.5 联合暴露实验

采用 96 h 静态暴露实验评估 TBBPA 的毒性机理及 PE-MPs 对其的影响。基于急性毒性实验的结果,采用 1/16 96 h-LC_{50} 值作为 TBBPA 的暴露浓度。选定 3 个环境相关的 PE-MPs 浓度(0.25 mg/L、5 mg/L 和 20 mg/L)作为暴露梯度[2-3],同时取各暴露溶液中的 DMSO 最高值设置 DMSO 溶剂对照。每个实验浓度和对照重复 3 次,其余条件与急性毒性实验相同。具体实验浓度设置见表 2-2。在联合暴露结束后,迅速对斑马鱼实施安乐死并用液氮进行速冻,随后转移至超低温冰箱中保存待测。

表 2-2 联合暴露实验暴露浓度

暴露组	TBBPA/(mg/L)	PE-MPs/(mg/L)	DMSO/(‰v/v)
对照组(CK)	0	0	0.005 3
TBBPA	0.053	0	0.005 3
TBBPA+0.25 PE	0.053	0.25	0.005 3
TBBPA+5 PE	0.053	5	0.005 3
TBBPA+20 PE	0.053	20	0.005 3

2.2.6 TBBPA 浓度测定

参照先前研究对暴露溶液中的 TBBPA 浓度进行测定[4]。在溶液配制完成后取 1.5 mL 待测,用 0.22 μm 滤膜对其进行过滤并转移至色谱进样瓶中。使用高效液相色谱仪(HPLC,1260 Infinity Ⅱ,Agilent,USA)测定溶液中 TBBPA 的实际浓度。色谱仪测定参数如下:C18 色谱柱,柱温 30 ℃,流动相为甲醇和水,体积比为 70∶30,进样体积为 1 μL,进样速度为 0.5 mL/min,利用可变波长检测器(VWD)在 210 nm 波长条件进行检测,TBBPA 的出峰时间为 2.1 min。

2.2.7 斑马鱼组织匀浆制备

从 CK、TBBPA、TBBPA+0.25 PE、TBBPA+5 PE、TBBPA+20 PE 组中随机选取 30 条斑马鱼(从每个烧杯中取 2 条,每组 3 次生物学重复)实施冰上安乐死,并立即将斑马鱼在 0.9% 生理盐水中冰浴条件下进行机械匀浆并制备成浓度为 10% 的组织匀浆,随后在 4 ℃、2 500 r/min 转速条件下离心 10 min 后,取上清液用于后续实验。

2.2.8 总蛋白(TP)含量测定

$$总蛋白浓度(mg/mL) = \frac{A_{测定} - A_{空白}}{A_{标准} - A_{空白}} \times 标准液浓度 \times 稀释倍数 \quad (2\text{-}1)$$

采用考马斯亮蓝法测定各样本中总蛋白含量以用于后续指标的归一化。基于预实验结果,采取浓度为 1% 的组织匀浆进行测定。依据表 2-3 进行实验操作,按顺序将待测液加入 96 孔板中(空白对照,标准对照与每个测定样本各进行 3 次技术重复),静置 10 min 后使用酶标仪以 595 nm 波长测定各孔吸光度(A),并利用式(2-1)计算总蛋白含量。

表 2-3 总蛋白含量测定操作表

	空白对照	标准对照	测定样本
超纯水/mL	0.05		
蛋白标准液/mL		0.05	
样品/mL			0.05
考马斯亮蓝显色液/mL	3.00	3.00	3.00

2.2.9 超氧化物歧化酶(SOD)活性测定

$$SOD\ 活性(U/mgport) = \frac{A_{对照} - A_{测定}}{A_{对照}} \div 2 \times \frac{反应液总体积/mL}{取样量/mL} \div 匀浆总蛋白浓度 \quad (2\text{-}2)$$

采用羟胺法测定各样本中 SOD 活性。基于预实验结果,采用浓度为 1% 的组织匀浆进行测定。依据表 2-4 进行实验操作,按顺序将待测液加入 96 孔板中(空白对照与每个测定样本各进行 3 次技术重复),混匀静置 10 min 后使用酶标仪以 550 nm 波长测定各孔吸光度(A),并利用式(2-2)计算 SOD 活性。

表 2-4 SOD 活性测定操作表

	空白对照	测定样本
试剂一/mL	1.00	1.00
样品/mL		0.05
超纯水/mL		0.05
试剂二/mL	0.10	0.10
试剂三/mL	0.10	0.10
试剂四应用液/mL	0.10	0.10
使用漩涡混匀器混匀,37 ℃恒温水浴 40 min		
显色剂/mL	2.00	2.00

2.2.10 过氧化氢酶(CAT)活性测定

$$CAT 活性(U/mgport) = \frac{(A_{对照} - A_{测定}) \times 271}{取样量/mL \times 反应时间/s \times 匀浆总蛋白浓度}$$
(2-3)

采用钼酸铵法测定各样本中 CAT 活性。基于预实验结果,采用浓度为 1% 的组织匀浆进行测定。依据表 2-5 进行实验操作,按顺序将待测液加入 96 孔板中(空白对照与每个测定样本各进行 3 次技术重复),后使用酶标仪以 405 nm 波长测定各孔吸光度(A),并利用式(2-3)计算 CAT 活性。

表 2-5 CAT 活性测定操作表

	空白对照	测定样本
样品/mL		0.05
试剂一(37 ℃)/mL	1.00	1.00
试剂二(37 ℃)/mL	0.10	0.10
混匀,37 ℃准确反应 60 s		
试剂三/mL	1.00	1.00
试剂四/mL	0.10	0.10
样品/mL	0.05	

2.2.11 丙二醛(MDA)含量测定

采用硫代巴比妥酸(TBA)法测定各样本中 MDA 含量。基于预实验结果，采取浓度为 10% 的组织匀浆进行测定。依据表 2-6 进行实验操作，随后将待测液 95 ℃ 水浴加热 40 min，冷却后以 3 500 r/min 转速条件下离心 10 min，随后按顺序将待测液加入 96 孔板中(空白对照、标准对照与每个测定样本各进行 3 次技术重复)，后使用酶标仪以 532 nm 波长测定各孔吸光度(A)，并利用式(2-4)计算 MDA 含量。

表 2-6 MDA 含量测定操作表

	空白对照	标准对照	测定样本	样本对照
样品/mL			0.10	0.10
10 nmol/L 标准品/mL		0.10		
无水乙醇/mL	0.10			
试剂一/mL	0.10	0.10	0.10	0.10
混匀				
试剂二/mL	3.00	3.00	3.00	3.00
试剂三/mL	1.00	1.00	1.00	
50%冰醋酸/mL				1.00

$$\text{MDA 含量(nmol/mL)} = \frac{A_{\text{测定}} - A_{\text{对照}}}{A_{\text{标准}} - A_{\text{空白}}} \times \text{标准品浓度} \div \text{匀浆总蛋白浓度}$$

(2-4)

2.2.12 组织病理学观察

选取 CK、TBBPA 和 TBBPA+5 PE 组的斑马鱼进行组织病理学观察。联合暴露实验结束后，将斑马鱼的肠道、鳃和肌肉立即固定在 4% 多聚甲醛溶液中。使用组织包埋机进行石蜡包埋，然后切成 3 μm 的切片并脱水。在经苏木精和伊红染色后，用全自动数码玻片扫描仪进行成像并观察。

2.2.13 统计学分析

数据以平均值±标准差(SD)表示。通过 SPSS 27(IBM, USA)进行单因素方差分析(ANOVA)，然后通过邓肯检验进行多重比较，以确定处理组之间差异的显著性。

2.3 TBBPA 的急性毒性

TBBPA 急性毒性实验结果如图 2-1 所示。

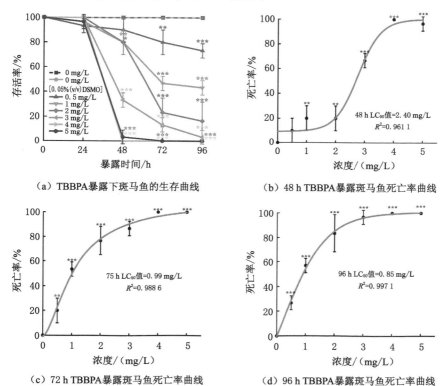

(a) TBBPA暴露下斑马鱼的生存曲线　　(b) 48 h TBBPA暴露斑马鱼死亡率曲线

(c) 72 h TBBPA暴露斑马鱼死亡率曲线　　(d) 96 h TBBPA暴露斑马鱼死亡率曲线

"*"表示暴露组斑马鱼与 0 h 时对比的生存率差异的显著性,*:$P<0.05$,* *:$P<0.005$,
* * *:$P<0.001$;通过 Probit 回归计算斑马鱼的 LC_{50} 值,结果以平均值±标准差表示($n=3$)。

图 2-1　TBBPA 对斑马鱼的急性毒性

实验期间空白对照组与溶剂对照组均未出现死亡。在暴露的第 48 h 开始,多数暴露组的斑马鱼开始出现显著的死亡现象,并且斑马鱼的致死效应随着 TBBPA 浓度的升高而加强。TBBPA 对斑马鱼的 48 h、72 h 及 96 h 的 LC_{50} 值分别为 2.40 mg/L、0.99 mg/L 和 0.85 mg/L。前人研究表明,双酚 A(BPA)对斑马鱼的 96 h LC_{50} 值为 12 mg/L,双酚 AF(BPAF)对斑马鱼的 96 h LC_{50} 值为 1.6 mg/L,双酚 F(BPF)对斑马鱼的 96 h LC_{50} 值为 32 mg/L,而双酚 S(BPS)对斑马鱼的 96 h LC_{50} 值为 199 mg/L[5]。这表明与这些塑料添加剂相比,TBBPA 具有更高的毒性。而对于其他水生物种,TBBPA 也表现出了不同的致死毒性。例如,TBBPA 对虹鳟鱼(*Oncorhynchus mykiss*)的 96 h LC_{50} 值为 1.1 mg/L[6],对胖头鲦鱼(*Pimephales promelas*)的 96 h LC_{50} 值为 0.54 mg/L[6],而对水蚤(*Acartia tonsa*)的 48 h LC_{50} 值为 0.4 mg/L[7]。结合上述研究结果,急性毒性实验表明 TBBPA 对斑马鱼等水生生物具有中度急性毒性。

2.4 PE-MPs 的粒径与形貌特征

PE-MPs 的表征结果如图 2-2 所示。扫描电镜的表征结果表明,PE-MPs 的粒径与标称尺寸一致。本研究中采用的 PE-MPs 表现出了不规则和粗糙的表面形态特征,这与环境中的 MPs 相似[8]。研究表明,粗糙的表面会增加 MPs 的比表面积,使得 MPs 表面对疏水性有机污染物的吸附位点增多,增强其对有机污染物的吸附能力[9]。对 TBBPA 的毒性影响更加明显。Yu 等的研究表明,两种从洗面奶中回收的粒径 100~400 μm 的 PE-MPs 对 TBBPA 的吸附能力为 87.26 mg/g 和 88.13 mg/g,这表明 PE-MPs 对 TBBPA 的毒性效应具有潜在影响能力。

 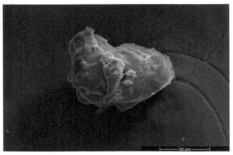

(a) PE-MPs的粒径分布(150×)　　　(b) 500×下PE-MPs的表面形貌和尺寸表征

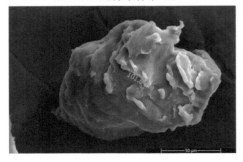 

(c) 500×下PE-MPs的表面形貌和尺寸表征　　　(d) 800×下PE-MPs的表面形貌和尺寸表征

图 2-2 PE-MPs 的扫描电子显微镜图像

2.5 暴露溶液中 TBBPA 浓度

HPLC 在空白对照组和 DMSO 溶剂对照组均未检测到 TBBPA(图 2-3)。急性毒性实验溶液中 TBBPA 的实测浓度分别为 0.52 mg/L、1.05 mg/L、2.05 mg/L、3.13 mg/L、3.81 mg/L 和 5.00 mg/L,浓度范围为标称浓度的 95.25%~104.33%。联合暴露溶液中 TBBPA 实测浓度分别为 53.36 μg/L、54.17 μg/L、53.64 μg/L 和 52.42 μg/L,实测浓度与标称浓度(53 μg/L)一致。HPLC 测定结果验证了暴露实验结果的可靠性,因此后续关于 TBBPA 毒性效应的讨论基于溶液标称浓度。

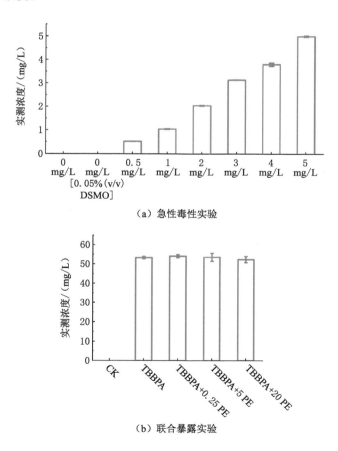

结果以平均值±标准差表示($n=3$)。

图 2-3 HPLC 对暴露溶液中 TBBPA 的浓度的定量

2.6 TBBPA 与 PE-MPs 对斑马鱼生化指标的影响

2.6.1 TBBPA 与 PE-MPs 对斑马鱼 SOD 活性的影响

联合暴露对斑马鱼 SOD 活性的影响如图 2-4 所示。结果表明,相比对照组,各实验组的 SOD 活性均被抑制。作为清除过量体内活性氧(ROS)的酶,SOD 的活性抑制机制可能是 TBBPA 诱导了斑马鱼的氧化损伤并诱导斑马鱼体内过量 ROS 的产生,而过量产生的 ROS 超出了 SOD 的催化能力,并氧化了蛋白的活性位点残基,导致 SOD 功能和活性的丧失[10]。与本研究结果类似,Shao 等[11]的研究表明,10 μg/L 硫丹暴露引起的氧化损伤抑制了斑马鱼的 SOD 活性。在复合暴露组中,除了最低浓度的 PE-MPs 外,其他浓度 PE-MPs 与 TBBPA 联合暴露均显著进一步抑制了 SOD 的活性。这表明 PE-MPs 作为新的应激源,显著加重了斑马鱼体内氧化损伤的水平,过量 ROS 的产生进一步抑制了 SOD 的活性。Zhang 等[12]的研究表明,聚苯乙烯微塑料(PS-MPs)共暴露显著增强了阿米替林对斑马鱼 SOD 活性的抑制作用,这支持了本研究的结果。

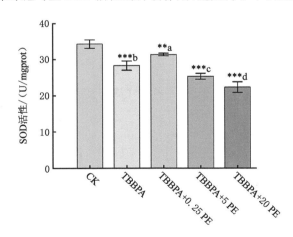

"*"表示处理组与对照组差异的显著性,*:$P<0.05$,* *:$P<0.01$,* * *:$P<0.001$;不同字母代表实验组的组间差异,结果以平均值±标准差表示($n=3$)。

图 2-4 联合暴露实验对斑马鱼 SOD 活性的影响

2.6.2 TBBPA 与 PE-MPs 对斑马鱼 CAT 活性的影响

联合暴露对斑马鱼 CAT 活性的影响如图 2-5 所示。与实验组对 SOD 活性的影响趋势不同,相比对照组,各实验组的 CAT 活性均被促进,TBBPA 单独暴露组的促进效应最为显著。而与 TBBPA 单独暴露组相比,随着共暴露组 PE-MPs 浓度的提高,CAT 活性受到显著抑制,且该抑制作用随着 PE-MPs 浓度上升而增强。CAT 作为催化 H_2O_2 分解成氧和水的酶,存在于斑马鱼的各个组织中,是保护机体免受氧化应激影响的重要抗氧化酶。TBBPA 对 CAT 活性的促进表明,TBBPA 诱导了斑马鱼过量的 ROS 产生,ROS 经抗氧化酶分解产生了大量 H_2O_2,导致机体 CAT 活性提高以降解 H_2O_2。而 PE-MPs 的加入导致了机体更严重的氧化损伤,H_2O_2 的过量产生超出了斑马鱼机体正常的降解范围,导致 CAT 的结构被氧化破坏进而导致其活性被抑制。

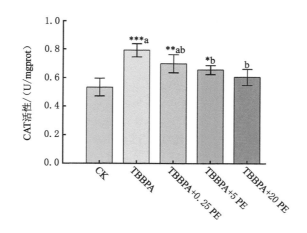

"*"表示处理组与对照组差异的显著性,*:$P<0.05$,* *:$P<0.01$,* * *:$P<0.001$;不同字母代表实验组的组间差异,结果以平均值±标准差表示($n=3$)。

图 2-5 联合暴露实验对斑马鱼 CAT 活性的影响

2.6.3 TBBPA 与 PE-MPs 对斑马鱼 MDA 含量的影响

联合暴露对斑马鱼 MDA 活性的影响如图 2-6 所示。相比对照组,TBBPA 暴露显著诱导了斑马鱼体内 MDA 含量的增加,而相比 TBBPA 暴露,TBBPA+0.25 PE 暴露组和 TBBPA+5 PE 暴露组的 MDA 含量没有显著增加,而 TBBPA+20 PE 暴露组则显著诱导了斑马鱼体内 MDA 含量的增加。MDA 是生物体内膜过氧化的产物,其含量的多少能直接反映生物氧化损伤水平。Feng 等[13]的研究表明,十溴二苯醚等溴代阻燃剂会诱导鲫鱼脂质过氧化,导致鲫鱼体内 MDA 含量异常升高;而 Jia 等[14]的研究表明,TBBPA 及其替代物四氯双酚 A(TCBPA)会造成青蛙体内的氧化损伤。结合本研究结果,TBBPA 会造成斑马鱼氧化损伤的产生,进而导致 MDA 含量的上升,而高浓度的 PE-MPs 会显著加强 TBBPA 诱导的氧化损伤,导致斑马鱼体内 MDA 含量的进一步提高。

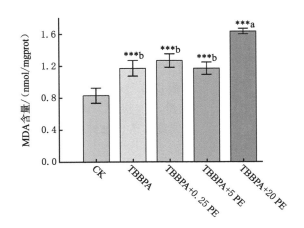

"*"表示处理组与对照组差异的显著性,*:$P<0.05$,**:$P<0.01$,***:$P<0.001$;不同字母代表实验组的组间差异,结果以平均值±标准差表示($n=3$)。

图 2-6 联合暴露实验对斑马鱼 MDA 活性的影响

2.7 TBBPA 与 PE-MPs 对斑马鱼组织的影响

对斑马鱼的组织病理学观察结果如图 2-7 所示。与对照组相比,TBBPA 暴露组和 TBBPA+5 PE 暴露组对斑马鱼的鳃和肌肉均无显著影响,而 TBBPA 暴露组和 TBBPA+5 PE 暴露组均对斑马鱼的肠道产生了显著影响,如肠道黏液分泌增加、纤毛和黏膜组织损伤和肠道管腔内组织碎片。相关研究表明,外源刺激诱导的免疫反应是斑马鱼肠道损伤的重要机制。肠道是鱼类消化、吸收和免疫的主要场所,在起吸收和输送营养物质作用的同时,还起抵御外界有害物质入侵的保护作用,外界污染物的刺激会诱导肠道组织产生过量的氧自由基攻击细胞膜并导致斑马鱼肠道的炎症等免疫反应,进而造成肠道的组织损伤[15]。例如,Sun 等[16]的研究表明,环康唑会诱导斑马鱼的肠道炎症反应,造成斑马鱼肠道上皮绒毛损伤、肠道壁变薄、杯状血细胞减少及炎症细胞浸润。Yu 等[17]的研究也表明,磷酸三苯酯暴露会诱导斑马鱼肠杯状细胞形态异常和肠绒毛溶解。因此,本研究中 TBBPA 暴露造成了斑马鱼的肠道免疫反应,进而导致了斑马鱼肠道损伤,而 PE-MPs 的加入作为共应激源加重了斑马鱼的肠道免疫反应。

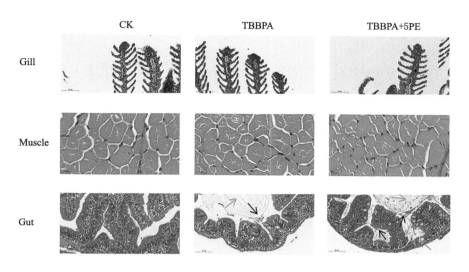

绿色箭头代表肠黏液分泌增加,黑色箭头代表纤毛和黏膜组织损伤,红色箭头代表肠腔内的组织碎片。

图 2-7 斑马鱼组织病理学分析结果

2.8 本章小结

为研究 TBBPA 对斑马鱼的毒性效应及 PE-MPs 对其的影响,本章以斑马鱼为受试生物,以 TBBPA 与不同浓度(0.25 mg/L、5 mg/L 和 20 mg/L)PE-MPs 为暴露污染物,通过急性毒性评估、生化指标测定及组织病理学观察等方法对 TBBPA 的毒性及 PE-MPs 对其的影响进行初步探究,结论如下:

① TBBPA 的 48 h、72 h 及 96 h 的 LC_{50} 值分别为 2.40 mg/L、0.99 mg/L 和 0.85 mg/L。

② 生化指标测定结果表明,TBBPA 暴露抑制了斑马鱼 SOD 活性并促进了 CAT 活性,同时诱导了 MDA 的过量产生,表明 TBBPA 暴露诱导了斑马鱼的氧化损伤。而 PE-MPs 存在情况下,TBBPA 对斑马鱼 SOD 活性的抑制加重,但 TBBPA 对斑马鱼 CAT 活性的促进作用受到抑制。此外,TBBPA+20 PE 暴露还导致了 MDA 的进一步过量产生,表明 PE-MPs 的存在加重了 TBBPA 诱导的斑马鱼的氧化损伤。

③ 组织病理学观察结果表明,PE-MPs 存在与不存在情况下,TBBPA 均未对斑马鱼的鳃和肌肉产生显著影响,但均诱导了斑马鱼肠道损伤,包括肠黏液分泌增加、纤毛和黏膜组织损伤和肠腔内的组织碎片增加等,同时 PE-MPs 存在明显加重了 TBBPA 对斑马鱼肠道的损伤程度。

参考文献

[1] OECD. Test No. 203: fish, acute toxicity test [M/OL]. https://max.book118.com/html/2017/0407/99228612.shtm.

[2] GREEN D S, BOOTS B, O'CONNOR N E, et al. Microplastics affect the ecological functioning of an important biogenic habitat[J]. Environmental science and technology, 2017, 51(1): 68-77.

[3] CUNNINGHAM E M, SIGWART J D. Environmentally accurate microplastic levels and their absence from exposure studies[J]. Integrative and comparative biology, 2019, 59(6): 1485-1496.

[4] YAO Y R, YIN L, HE C, et al. Removal kinetics and mechanisms of tetrabromobisphenol A (TBBPA) by HA-n-FeS colloids in the absence and presence of oxygen[J]. Journal of environmental management, 2022, 311: 114885.

[5] MOREMAN J, LEE O, TRZNADEL M, et al. Acute toxicity, teratogenic, and estrogenic effects of bisphenol A and its alternative replacements bisphenol S, bisphenol F, and bisphenol AF in zebrafish embryo-larvae[J]. Environmental science and technology, 2017, 51(21): 12796-12805.

[6] PITTINGER C A, PECQUET A M. Review of historical aquatic toxicity and bioconcentration data for the brominated flame retardant tetrabromobisphenol A (TBBPA): effects to fish, invertebrates, algae, and microbial communities[J]. Environmental science and pollution research, 2018, 25(15): 14361-14372.

[7] WOLLENBERGER L, DINAN L, BREITHOLTZ M. Brominated flame retardants: activities in a crustacean development test and in an ecdysteroid screening assay[J]. Environmental toxicology and chemistry, 2005, 24(2): 400-407.

[8] XIA F Y, TAN Q, QIN H G, et al. Sequestration and export of microplastics in urban river sediments[J]. Environment international, 2023, 181: 108265.

[9] FU L N, LI J, WANG G Y, et al. Adsorption behavior of organic pollutants on microplastics[J]. Ecotoxicology and environmental safety, 2021, 217: 112207.

[10] BUTTERFIELD D A, KOPPAL T, HOWARD B, et al. Structural and functional changes in proteins induced by free radical-mediated oxidative stress and protective action of the antioxidants N-tert-butyl-alpha-pheny-

lnitrone and vitamin E[J].Annals of the New York academy of sciences,1998,854:448-462.

[11] SHAO B,ZHU L S,DONG M,et al.DNA damage and oxidative stress induced by endosulfan exposure in zebrafish (Danio rerio)[J].Ecotoxicology,2012,21(5):1533-1540.

[12] ZHANG Y,CHEN C,CHEN K.Combined exposure to microplastics andamitriptyline induced abnormal behavioral responses and oxidative stress in the eyes of zebrafish (Danio rerio)[J].Comparative biochemistry and physiology toxicology and pharmacology,2023,273:109717.

[13] FENG M B,QU R J,WANG C,et al.Comparative antioxidant status in freshwater fish Carassius auratus exposed to six current-use brominated flame retardants:a combined experimental and theoretical study[J].Aquatic toxicology,2013,140/141:314-323.

[14] JIA X Y,YAN R P,LIN H K,et al.TBBPA and its alternative TCBPA induced ROS-dependent mitochondria-mediated apoptosis in the liver of Rana nigromaculata[J].Environmental pollution,2022,297:118791.

[15] HUANG T,SUN D Y,YANG W X,et al.The fabrication of porous Si with interconnected micro-sized dendrites and tunable morphology through the dealloying of a laser remelted Al-Si alloy[J].Materials,2017,10(4):357.

[16] SUN X X,TIAN S N,YAN S,et al.Bifidobacterium mediate gut microbiota-remedied intestinal barrier damage caused by cyproconazole in zebrafish (Danio rerio)[J].Science of the total environment,2024,912:169556.

[17] YU F R,LIU Y H,WANG W Y,et al.Toxicity of TPhP on the gills and intestines of zebrafish from the perspectives of histopathology, oxidative stress and immune response[J].Science of the total environment,2024,908:168212.

第3章 聚乙烯微塑料对四溴双酚A斑马鱼毒性影响机制

3.1 引言

生物信息学技术从分子层面研究环境污染物对生物的毒性效应,是揭示环境污染物毒理机制的有效手段。本章基于转录组测序技术,结合生物信息学方法,分析了 TBBPA 对斑马鱼毒性效应的作用机制,以及 PE-MPs 存在时对 TBBPA 毒性效应的影响。同时,借助实时荧光定量 PCR、分子对接和分子动力学模拟等方法,进一步验证了 TBBPA 对斑马鱼的毒性机制及 PE-MPs 的影响机理。

3.2 实验材料与方法

3.2.1 受试生物

受试生物同 2.2.1。

3.2.2 试剂与仪器

四溴双酚 A(TBBPA,$C_{15}H_{12}Br_4O_2$,CAS 79-94-7,纯度≥99%)和二甲基亚砜(DMSO,CAS 67-68-5,纯度≥99.9%)购自上海阿拉丁生化科技股份有限公司。100 μm PE-MPs(纯度≥99%)购自上海阳励机电科技有限公司。实验前将 TBBPA 溶于 DMSO 中制成浓度为 10 g/L 的 TBBPA 储备母液,并于 −18 ℃条件下储存待用。

本章所用实验仪器见表 3-1。

表 3-1 实验仪器

仪器名称	仪器型号	生产厂家
超声波清洗器	KH-500DE	昆山禾创超声仪器有限公司
冰箱	BCD-452WDPF	青岛海尔股份有限公司
超低温冰箱	DW-86L100	青岛海尔生物医疗股份有限公司
超纯水系统	OKP-10	上海涞科仪器有限公司
电子天平	FA2204N	上海菁海仪器有限公司
高压灭菌器	MLS-378K-PC	Panasonic
水质检测仪	HI9829-04	Hanna Instruments
智能人工气候箱	RXZ-1500c-2	宁波江南仪器厂
转录组测序平台	Illumina Novaseq 6000	Illumina
RT-qPCR 系统	ThermoFisher	Illumina

3.2.3 转录组测序与差异表达基因鉴定

使用 TRIzol 试剂（Invitrogen, USA）提取联合暴露实验中各组斑马鱼的总 RNA，随后使用生物分析仪（Agilent2100, Agilent）进行总 RNA 质量控制（QC）。选定高质量 RNA 样品（RIN≥7, 28S/18S≥1.5）用于后续构建文库。使用 Qubit RNA 检测试剂盒（Invitrogen, USA）对总 RNA 进行定量，随后将总 RNA 中的 mRNA 进行纯化和片段化用于制备 RNA-Seq 文库，并在转录组测序平台（Illumina Novaseq 6000）上进行测序。在评估所有 mRNA 的表达水平后，使用 limma R 包进行差异表达分析。以差异表达倍数（fold change, FC）定义差异表达基因。$\log_2 FC \geq 1$ 或 ≤ -1 且 $P < 0.05$ 的基因被定义为差异表达基因。

3.2.4 GO 富集分析

GO 是基因本体联合会所建立的语义词汇标准，旨在建立一个适用于各种物种的、能随着研究进展不断更新的、对基因和蛋白质功能进行限定和描述的数据库。GO 功能显著性富集分析的目的是根据目标基因集合与背景基因集合的匹配程度分析目标基因集合与哪些生物学功能显著相关。本研究将转录组测序数据向 GO 数据库（http://www.geneontology.org/）的各个条目映射，计算每个条目的基因数目，利用超几何分布进行假设检验得到富集结果的显著性，P 值越低富集结果越显著。随后，利用 Cytoscape 软件的 ClueGO 插件进行网络分析及可视化。

3.2.5 KEGG 富集分析

京都基因与基因组百科全书(KEGG)数据库是从分子水平信息了解高级功能和生物系统,尤其是大型分子数据集生成的基因组测序和其他高通量实验技术的实用程序数据库资源。KEGG 富集分析整合了当前分子互动网络的知识,有助于从生理功能层面揭示化学品暴露对斑马鱼的影响[1]。本研究利用 KOBAS 工具,将转录组测序数据导入 KEGG 数据库(https://www.genome.jp/kegg/)的各个通路映射,并计算每个通路的基因数目,利用超几何分布进行假设检验得到富集结果的显著性,P 值越低富集结果越显著。随后,利用 Omicshare 生物信息学分析平台(https://www.omicshare.com/)进行通路网络分析。

3.2.6 Reactome 富集分析

Reactome 是一个开源数据库,由生物学专家与 Reactome 编辑人员合作创建通路以及注释,其数据模型概括了反应的概念,包括小分子的运输、复合物的相互作用以及经典生物化学的化学转换等。该数据库的功能是对生物生理过程的详细描述,包括信号传导、代谢、转录调节、细胞凋亡和突触传递等过程[2]。本研究利用 KOBAS 工具,将转录组测序数据导入 Reactome 数据库(https://reactome.org/)的各个通路映射,并计算每个通路的基因数目,利用超几何分布进行假设检验得到富集结果的显著性,P 值越低富集结果越显著。

3.2.7 Panther 富集分析

Panther 是一个基因家族数据库,包括每个家族的系统发育树,树的节点都标有基因属性。目前在 Panther 中注释的基因属性有三种类型,即"亚家族成员""蛋白质类别""基因功能"。其内部节点通过其所代表的进化事件类型进行注释,如物种形成、基因复制或水平基因转移等事件[3]。本研究利用 KOBAS 工具,将转录组测序数据导入 Panther 数据库(http://www.pantherdb.org/)的各个通路映射,并计算每个通路的基因数目,利用超几何分布进行假设检验得到富集结果的显著性,P 值越低富集结果越显著。

3.2.8 基因集富集分析

基因集富集分析(GSEA)使用预定义的基因,将基因按照在两类样本中的差异表达程度排序,然后检验预先设定的基因集合在这个排序表中是否具有整体上调或下调的趋势。其相比于 GO 分析和通路分析的优势在于,后两者会提前设定一个阈值,只关注差异变化大的基因。GSEA 整体不预先设定分析阈值,不易遗漏部分差异表达不显著却有重要生物学意义的基因。本研究利用 Omicshare 生物信息学分析平台(https://www.omicshare.com/)对相关暴露组的基因进行 GSEA,以确定暴露引起的基因在相关通路中的调节趋势,利用超几何分布进行假设检验得到富集结果的显著性,P 值越低富集结果越显著。

3.2.9 蛋白互作网络分析

蛋白互作网络(protein-protein interaction network, PPI)分析是描述一组蛋白间相互作用关系的分析方式。PPI 分析的目的是明确差异表达基因编码的蛋白之间的相互作用关系,同时还能筛选出一些处于蛋白互作网络上关键节点的蛋白。本研究利用 STRING 在线工具(https://cn.string-db.org),将选定通路内的差异表达基因名称导入数据库中并将物种设置为斑马鱼,以 0.7 高等可信度进行 PPI 网络的构建。在网络构建完成后利用 Cytoscape 软件的 cytoHubba 插件进行网络的可视化。

3.2.10 实时荧光定量 PCR(RT-qPCR)

使用 TRNzol 通用试剂提取样本中的总 RNA。在纯化后,使用逆转录试剂盒(艾科瑞生物,AG11728,中国)对总 RNA 进行逆转录。随后,使用 RT-qPCR 系统(QuantStudio 7 Flex,ThermoFisher,USA)进行 RT-qPCR。RT-qPCR 的引物、反应体系和程序见表 3-2。采用 $2^{-\Delta\Delta Ct}$ 法计算 mRNA 的相对表达量,并与内参基因 β-actin 进行标准化,见表 3-2~表 3-4。

表 3-2 RT-qPCR 引物

基因	Ensembl ID	方向	Sequence (5'-3')	长度/bp
*prim*2	ENSDARG00000052721	Forward	AACGTTATCTCTGTTGTGTAGGTT	123
		Reverse	TCCAGAGGAGGCTGACCATA	
thrb	ENSDARG00000021163	Forward	GACAAATGCAACTCCCGCTG	148
		Reverse	AACCTTTGCAGCCCTCACAT	
*itga*11a	ENSDARG00000036086	Forward	AGAGGGGACAAGCAAGAACG	154
		Reverse	CACTTTCCCTTGCCGAGTCT	
*cyp*17a1	ENSDARG00000033566	Forward	TCAGCGACAGGGGGAATCTA	77
		Reverse	GAGCGGAGAAACAGGTCGAA	
*prdx*1	ENSDARG00000058734	Forward	GTGGCAGATACTCTCCGCTC	137
		Reverse	CCGACTGGCAGATCGTTGAT	

表 3-2(续)

基因	Ensembl ID	方向	Sequence (5'-3')	长度/bp
nphs2	ENSDARG00000042850	Forward	CCTGACCCAATCCATCGACC	86
		Reverse	TGGTTCCATCATCTGAGCCG	
klf3	ENSDARG00000015495	Forward	CCTTCGTATCGAGTCCTGGC	179
		Reverse	AAGCTCATGGTGGTCCGAAG	
tmprss15	ENSDARG00000079393	Forward	CCTGGTTCCTGCGAACAGAT	115
		Reverse	AGCTGAACGTCGAGTTTGGT	
β-actin	ENSDARG00000037746	Forward	GCATTGCTGACCGTATGCAG	162
		Reverse	ACTCCTGCTTGCTGATCCAC	

表 3-3 RT-qPCR 的反应体系

反应体系	体积
2×SG Green qPCR Mix	5 μL
Forward Primer (10 μm)	0.2 μL
Reverse Primer (10 μm)	0.2 μL
cDNA	1 μL
Water, nuclease-free	3.6 μL
Total volume	10 μL

表 3-4 RT-qPCR 的反应程序

步骤	温度/℃	时间	循环数
Initial denaturation	95	3 min	1
Denaturation	95	10 s	40
Annealing+extension	60	30 s	

3.2.11 分子对接

斑马鱼甲状腺激素受体 β(THRβ,Uniport ID:Q9PVE4)和 17β-羟基类固醇脱氢酶 1(HSD17β1,Uniport ID:F1QUP2)的受体蛋白 3D 结构下载于 Uniport 数据库(https://www.uniprot.org/),TBBPA(CID:6618)、三碘甲状腺原氨酸(T_3,化合物 CID:5920)、四碘甲状腺原氨酸(T_4,CID:5819)、雌酮(E_1,CID:5870)和雌二醇(E_2,CID:5757)的 3D 结构下载于 Pubchem 数据库(https://

pubchem.ncbi.nlm.nih.gov)。将配体和受体蛋白导入 Discovery Studio 软件(BIOVIA,France)中,在 Receptor-Ligand Interactions 模块中获得受体蛋白的结合腔。随后使用 Attributes 工具设置结合腔的结合半径。然后,利用 Dock-Ligands 模块对受体蛋白与配体进行对接模拟。

3.2.12 分子动力学模拟

利用 GROMACS 5.1.4 软件,采用蛙跳牛顿积分法模拟配体与受体蛋白的结合复合物。采用分子力学/泊松-玻尔兹曼表面积(MMPBSA)方法进行 30 ns 分子动力学模拟,基于动力学稳定后的平均结构,分别计算复合物(G_{complex})、受体($G_{\text{free-protein}}$)和配体($G_{\text{free-ligand}}$)的结合自由能并计算总结合自由能(G_{bind})。计算公式如下:

$$G_{\text{bind}} = G_{\text{complex}} - G_{\text{free-protein}} - G_{\text{free-ligand}} \tag{3-1}$$

3.2.13 统计学分析

生物信息学分析结果采用超几何分布进行假设检验来验证分析结果的显著性,最终结果均取 $P<0.05$ 为显著阈值。

3.3 转录组测序实验质量控制与结果

3.3.1 转录组测序实验质量控制

转录组测序的原始文件通过 bcl2fastq 软件处理得到原始序列,并经过质量控制去除不满足分析标准的序列得到过滤后的序列,结果见表 3-5。转录组测序共得到 670 256 588 条原始序列,过滤后序列数量为 657 816 520 条,最终处理得到的总测序有效序列占总序列的比例为 97.29%,且所有样本中的有效序列比例均高于 95%。同时,各样品的测序质量值高于 20 的碱基数占总碱基数的百分比(Q20)均高于 98.5%,测序质量值高于 30 的碱基数占总碱基数的百分比(Q30)均高于 95.0%,这表明转录组测序结果具有较高的准确性和可靠性,确保了后续生物信息学分析的准确性。

表 3-5 转录组测序及质量控制数据统计

样本	原始序列数	原始序列总碱基数	过滤后序列数	过滤后序列总碱基数	有效序列/%	Q20/%	Q30/%
CK1	40 937 308	6 140 596 200	40 204 878	5 978 256 246	97.36	98.77	95.76

表 3-5(续)

样本	原始序列数	原始序列总碱基数	过滤后序列数	过滤后序列总碱基数	有效序列/%	Q20/%	Q30/%
CK2	44 594 486	6 689 172 900	43 845 050	6 520 497 413	97.48	98.83	95.93
CK3	48 979 104	7 346 865 600	47 966 284	7 141 097 167	97.20	98.51	95.21
TB1	44 350 786	6 652 617 900	43 527 690	6 465 437 793	97.19	98.84	96.00
TB2	45 288 912	6 793 336 800	44 445 618	6 602 588 681	97.19	98.75	95.71
TB3	41 637 436	6 245 615 400	40 893 326	6 074 140 977	97.25	98.79	95.83
TB+0.25 PE1	47 126 294	7 068 944 100	46 321 838	6 893 958 453	97.52	98.61	95.32
TB+0.25 PE2	47 347 406	7 102 110 900	46 614 820	6 941 501 356	97.74	98.62	95.30
TB+0.25 PE3	43 633 218	6 544 982 700	42 979 024	6 400 751 211	97.80	98.68	95.50
TB+5 PE1	40 253 902	6 038 085 300	39 667 680	5 907 843 705	97.84	98.73	95.63
TB+5 PE2	45 134 570	6 770 185 500	44 523 892	6 628 033 953	97.90	98.80	95.84
TB+5 PE3	49 285 626	7 392 843 900	48 143 044	7 135 926 046	96.52	98.82	95.88
TB+20 PE1	43 700 354	6 555 053 100	42 185 004	6 239 274 242	95.18	98.67	95.45
TB+20 PE2	42 359 120	6 353 868 000	41 670 262	6 207 644 574	97.70	98.81	95.89
TB+20 PE3	45 628 066	6 844 209 900	44 828 110	6 669 276 588	97.44	98.71	95.62

3.3.2　差异表达基因鉴定结果

利用转录组测序方法鉴定联合暴露实验中不同暴露条件诱导的差异表达基因,结果如图 3-1 所示。相对于对照组,TBBPA 暴露组诱导了 1 341 个基因显著差异表达,其中上调基因 941 个、下调基因 400 个[图 3-1(a)];TBBPA+0.25 PE 暴露组诱导了 1 835 个基因显著差异表达,其中上调基因 1 201 个、下调基因 634 个[图 3-1(b)];TBBPA+5 PE 暴露组诱导了 758 个基因显著差异表达,其中上调基因 507 个、下调基因 251 个[图 3-1(c)];TBBPA+20 PE 暴露组诱导了 554 个基因显著差异表达,其中上调基因 429 个、下调基因 125 个[图 3-1(d)]。在所有暴露组中被显著上调的基因均高于被显著下调的基因,也即 TBBPA 单独暴露及其与 PE-MPs 联合暴露诱导的上调表达基因均多于下调表达基因,表明 TBBPA 主要通过显著诱导基因上调对斑马鱼造成毒理影响。在差异表达基因的分布上,TBBPA 暴露组和 TBBPA+0.25 PE 暴露组共享了最多的差异表达基因,共 934 个,分别占两组差异表达基因总数的 69.6% 和 50.9%[图 3-1(e)],这表明低浓度 PE-MPs 的加入对 TBBPA 的毒理影响较小,而随着 PE-MPs 的浓度提高,其对 TBBPA 的毒性影响也逐渐增强。采用主成分分析(PCA)方法分析不同暴露组基因差异表达模式的差异性[图3-1(f)],分析结果解释了数据总方差的52.9%。由样本得分图可以看出,4个暴露组的样本差异表达基因的分组模式明显不同。对照组样品主要分布在PC1轴的负值区和PC2轴的正值区,而4个暴露组的分布模式则与对照组明显不同。此外,4个暴露组之间也存在明显的分区。这证实了TBBPA暴露会引起斑马鱼转录组的畸变,同时PE-MPs会对这种效应产生影响。

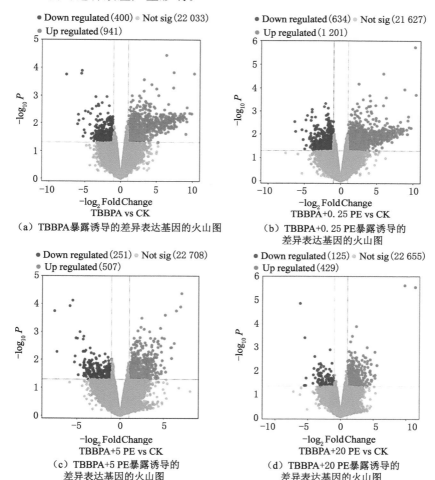

（a）TBBPA暴露诱导的差异表达基因的火山图

（b）TBBPA+0.25 PE暴露诱导的差异表达基因的火山图

（c）TBBPA+5 PE暴露诱导的差异表达基因的火山图

（d）TBBPA+20 PE暴露诱导的差异表达基因的火山图

横坐标为基因差异表达的程度,以差异表达倍数的对数表示;纵坐标为基因差异表达显著性,以 P 值的对数表示。蓝色和红色圆点分别表示显著下调和上调的基因,灰色圆点表示不显著的差异表达基因。

图3-1 利用转录组测序方法鉴定联合暴露实验中不同暴露条件诱导的差异表达基因结果

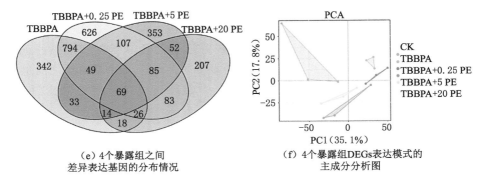

(e) 4个暴露组之间差异表达基因的分布情况　　(f) 4个暴露组DEGs表达模式的主成分分析图

图 3-1 （续）

3.4　TBBPA 对斑马鱼毒性影响的潜在机理

3.4.1　TBBPA 诱导的斑马鱼差异表达基因的 GO 富集结果

TBBPA 暴露诱导的斑马鱼差异表达基因的 GO 二级功能分类统计结果如图 3-2 所示。统计结果表明，TBBPA 暴露诱导的斑马鱼差异表达基因显著富集于细胞过程（上调 455，下调 165）、单生物过程（上调 355，下调 160）、代谢过程（上调 320，下调 91）以及生物调节（上调 240，下调 107）等生物调节过程，作用于结合（上调 448，下调 159）、催化活性（上调 205，下调 71）、分子功能调节（上调 33，下调 10）、转运活性（上调 26，下调 25）等分子功能，影响细胞（上调 387，下调 120）、细胞组分（上调 387，下调 120）、细胞器（上调 303，下调 72）、细胞器组分（上调 162，下调 33）等细胞组分。

分别从生物过程、细胞组分和分子功能三个类别统计 TBBPA 暴露诱导的斑马鱼差异表达基因富集的显著性最高的十个 GO 条目，结果见表 3-6。富集最显著的生物过程条目与细胞周期和生殖相关，包括单生物生殖过程（GO：0044702）、细胞周期（GO：0007049）、生殖过程（GO：0022414）等生物过程；富集最显著的细胞组分条目与染色体相关，包括极质（GO：0045495）、染色体（GO：0005694）、染色体部分（GO：0044427）等细胞组分；富集最显著的分子功能条目与 DNA 的复制相关，包括顶体酶结合（GO：0032190）、解旋酶活性（GO：0004386）、核质转运体活性（GO：0005487）等分子功能。GO 条目的富集结果表明，TBBPA 诱导的差异表达基因大量富集于细胞周期、细胞复制、染色体、生殖细胞生成过程等条目，这意味着 TBBPA 对斑马鱼产生了明显的遗传毒性影响。

第 3 章 聚乙烯微塑料对四溴双酚 A 斑马鱼毒性影响机制

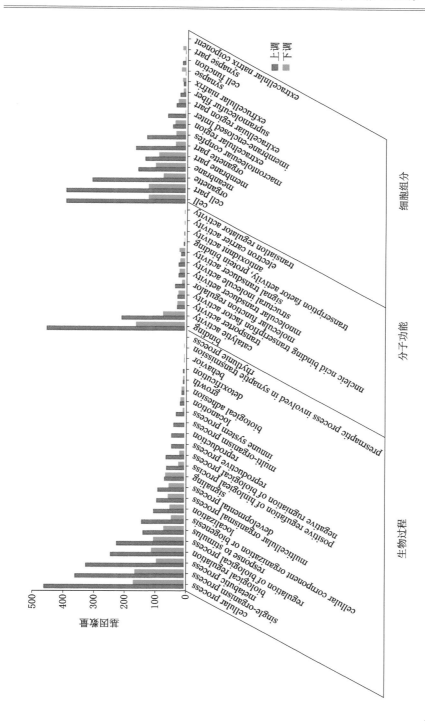

图3-2 TBBPA暴露诱导的斑马鱼差异表达基因的GO二级功能分类统计

表 3-6 TBBPA 暴露诱导的差异表达基因的 GO 富集结果

GO Term	Database	ID	P 值
single organism reproductive process	生物过程	GO:0044702	2.462×10^{-16}
cell cycle	生物过程	GO:0007049	5.017×10^{-16}
reproductive process	生物过程	GO:0022414	8.069×10^{-15}
reproduction	生物过程	GO:0000003	1.075×10^{-15}
DNA metabolic process	生物过程	GO:0006259	6.407×10^{-15}
cell cycle process	生物过程	GO:0022402	5.160×10^{-14}
DNA replication	生物过程	GO:0006260	5.474×10^{-14}
mitotic cell cycle	生物过程	GO:0000278	5.127×10^{-13}
cellular process involved in reproduction in multicellular organism	生物过程	GO:0022412	6.656×10^{-13}
oogenesis	生物过程	GO:0048477	1.631×10^{-12}
pole plasm	细胞组分	GO:0045495	1.029×10^{-10}
chromosome	细胞组分	GO:0005694	1.916×10^{-10}
chromosomal part	细胞组分	GO:0044427	1.965×10^{-10}
non-membran10-bounded organelle	细胞组分	GO:0043228	1.584×10^{-9}
intracellular non-membran10-bounded organelle	细胞组分	GO:0043232	1.584×10^{-9}
P granule	细胞组分	GO:0043186	1.690×10^{-9}
germ plasm	细胞组分	GO:0060293	1.690×10^{-9}
replication fork	细胞组分	GO:0005657	5.629×10^{-7}
chromosome, centromeric region	细胞组分	GO:0000775	3.276×10^{-6}
ribonucleoprotein granule	细胞组分	GO:0035770	4.433×10^{-6}
acrosin binding	分子功能	GO:0032190	4.214×10^{-7}
helicase activity	分子功能	GO:0004386	0.000 560 38
nucleocytoplasmic transporter activity	分子功能	GO:0005487	0.000 668 87
carbohydrate binding	分子功能	GO:0030246	0.000 816 17
mRNA 3′-UTR binding	分子功能	GO:0003730	0.001 070 34
nuclear localization sequence binding	分子功能	GO:0008139	0.001 070 34
palmitoyl hydrolase activity	分子功能	GO:0098599	0.001 197 36
palmitoyl-(protein) hydrolase activity	分子功能	GO:0008474	0.001 197 36
DNA helicase activity	分子功能	GO:0003678	0.001 217 84
fatty acid synthase activity	分子功能	GO:0004312	0.001 615 01

利用 Cytoscape 软件的 ClueGO 插件对 GO 条目进行网络分析,结果如图 3-3 所示。TBBPA 暴露组的主要功能网络包括细胞周期过程,DNA 复制,P 颗粒、有丝分裂核分裂、核染色体生成、染色体、卵子生成、雌性配子生成,生殖过程的调控等网络,这进一步表明了 TBBPA 暴露会对斑马鱼造成显著的遗传毒性效应。

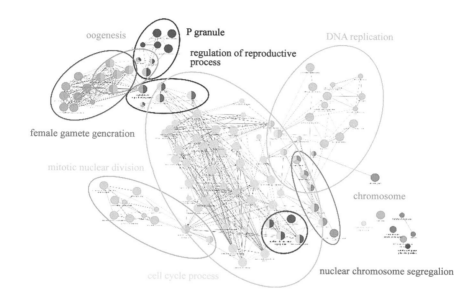

图 3-3　TBBPA 暴露诱导差异表达基因的 GO 富集网络分析

3.4.2　TBBPA 暴露诱导斑马鱼差异表达基因的通路富集结果

将 TBBPA 暴露诱导斑马鱼差异表达基因 KEGG、Reactome、Panther 三个数据库进行通路富集分析,以进一步探究 TBBPA 暴露对斑马鱼的生物学影响,结果如图 3-4(a)所示。TBBPA 暴露诱导的差异表达基因最显著富集的通路均与细胞周期与遗传相关,如细胞周期通路、DNA 复制通路、卵母细胞减数分裂通路等,与 GO 分析结果相符,表明 TBBPA 对斑马鱼造成了显著的遗传毒性影响。随后,利用 KEGG 数据库对差异表达基因进行功能注释[图 3-4(b)],结果表明 TBBPA 暴露对斑马鱼的细胞生长与死亡、转录、复制与修复、内分泌系统与免疫系统产生了影响。同时对 TBBPA 诱导差异表达基因富集的 KEGG 通路进行网络分析[图 3-5(a)],结果表明 TBBPA 通过类固醇激素的生物合成间接影响了卵母细胞的生成与分裂,此外还通过细胞周期-p53 信号通路-DNA 复

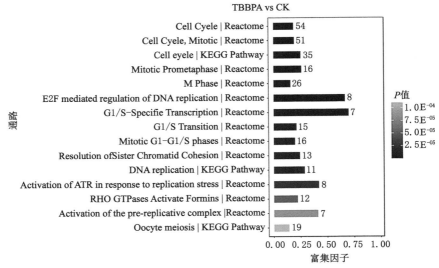

(a) TBBPA暴露诱导差异表达基因的通路富集分析

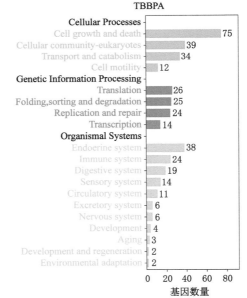

(b) 在KEGG通路富集的差异表达基因的二级功能注释

横坐标表示注释的差异基因占该通路中所有基因的频率,纵坐标为各数据库以显著性排序最显著富集的15个通路;数字代表富集基因的数量;颜色代表显著性,不同颜色代表不同的功能分类。

图3-4 TBBPA暴露诱导差异表达基因的通路富集分析及KEGG通路富集的差异表达基因的二级功能注释

第 3 章 聚乙烯微塑料对四溴双酚 A 斑马鱼毒性影响机制

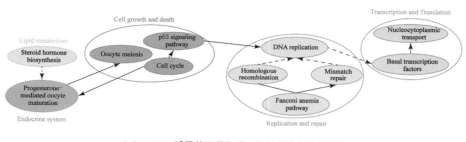

(a) TBBPA诱导的斑马鱼遗传相关通路扰动的网络

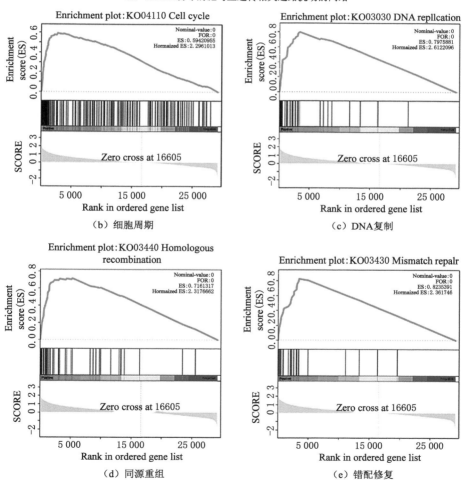

(b) 细胞周期　　　　　　　　　　(c) DNA复制

(d) 同源重组　　　　　　　　　　(e) 错配修复

GSEA 分析结果验证了 KEGG 通路的调节趋势；P 值为 0 表明实测的 $P<0.01$。

图 3-5　TBBPA 诱导的斑马鱼遗传相关通路扰动的网络及 GSEA 分析结果

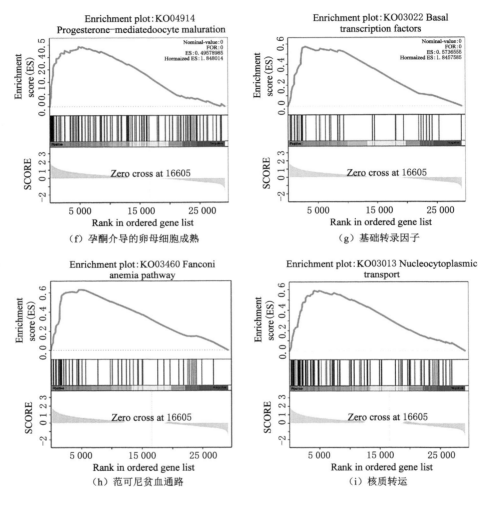

图 3-5 （续）

制途径影响斑马鱼细胞的正常分裂,同时,范可尼贫血通路等 DNA 复制修复通路被激活,表明斑马鱼的染色体复制出现显著异常,进而影响到了遗传信息的转录与翻译。采用 GSEA 验证差异表达基因对相关 KEGG 通路的调节趋势进行验证[图 3-5(b)～(i)],细胞周期、DNA 复制、同源重组、错配修复、孕酮介导的卵母细胞成熟、基础转录因子、范可尼贫血通路和核质转运均被验证为显著上调 ($P<0.01$),进一步验证了 TBBPA 的遗传毒性影响。

3.4.3　TBBPA 暴露对斑马鱼细胞周期和 DNA 复制的影响

细胞周期是细胞中发生的一系列事件，其作用是调控生物的基因组复制和染色体分裂并维持基因组完整性[4]。通路分析表明，TBBPA 暴露导致编码细胞周期蛋白依赖性激酶（CDKs）和细胞周期蛋白（Cycs）相关基因的表达上调[图 3-6(a)]。CDKs 在真核细胞分裂周期的协调中发挥整合生长调节信号的重要作用[5]。CDKs 可以分为两个主要子类，CDK1、CDK2、CDK4 和 CDK6 直接调控细胞周期进程，而 CDK7 和 CDK9 通常通过调控转录间接调控细胞周期进程[6]。Cycs 在细胞周期中期性地积累和降解[7]。根据结构相似性和周期性积累的模式，Cycs 可分为 G1 特异性细胞周期蛋白（CCNCs、CCNDs 和 CCNEs）和有丝分裂周期细胞周期蛋白（CCNAs 和 CCNBs）。每个 Cycs 可以与细胞中的几个 CDKs 相互作用形成 Cyc-Cdk 复合物[8]。Cyc-Cdk 复合物是调控细胞周期的关键分子，通过调控特定靶点的磷酸化来驱动细胞周期[9]，同时作为细胞周期引擎负责协调细胞生长、DNA 复制和确保后代细胞的活力[10]。CDKs 和 Cycs 相关基因的异常表达可导致真核细胞分裂过程中的各种问题。研究表明，编码 CDKs 基因的过表达可导致细胞周期速率增加、细胞增殖以及基因组和染色体不稳定[11-12]。而编码 CycA 基因的过表达会导致细胞过早进入有丝分裂 S 期、有丝分裂 S 期延长和染色体损伤，如双链断裂和中心体过表达[13-15]。采用 PPI 分析研究在 TBBPA 对斑马鱼细胞周期影响作用中的关键蛋白，结果如图 3-6(b)所示。在蛋白互作网络中以中心度排序，排名前四的蛋白分别是细胞周期蛋白 B1（CCNB1）、细胞周期蛋白依赖性激酶 2（CDK2）、细胞分裂周期蛋白 20（CDC20）和有丝分裂停滞缺陷样 1 结合蛋白（MAD2L1）。其中，CCNB1 和 CDK2 是编码 Cyc-Cdk 复合物的相关蛋白。CDC20 和 MAD2L1 分别参与细胞有丝分裂的 G2 期和 M 期。这些蛋白参与了细胞周期中多种复杂过程，在 TBBPA 诱导的遗传毒性中可能起关键作用。

先前的体外研究表明，低浓度的 TBBPA 可以加速 HepG2 细胞的增殖，影响细胞分裂后的遗传一致性[16]。在本研究中，TBBPA 诱导了大量细胞周期调控的相关基因上调。细胞周期调控基因的异常表达可能与癌症、神经系统疾病和心血管疾病等多种疾病的病理生理学有关，细胞周期紊乱导致细胞周期中终止染色体异常复制的能力降低被认为是癌症的诱导机理之一[17]。例如，*cdc45* 的显著上调可能是甲状腺癌等多种癌症的诱因，*cdc20* 的过表达也是肺癌发生的标志[18]。对于神经系统而言，细胞周期的异常与胶质瘢痕的形成和炎性因子的产生有关，并会导致中枢神经系统的急性损伤和慢性退行性疾病[19]。还有研究证实，细胞周期失控也是诱导中枢神经系统神经元死亡的

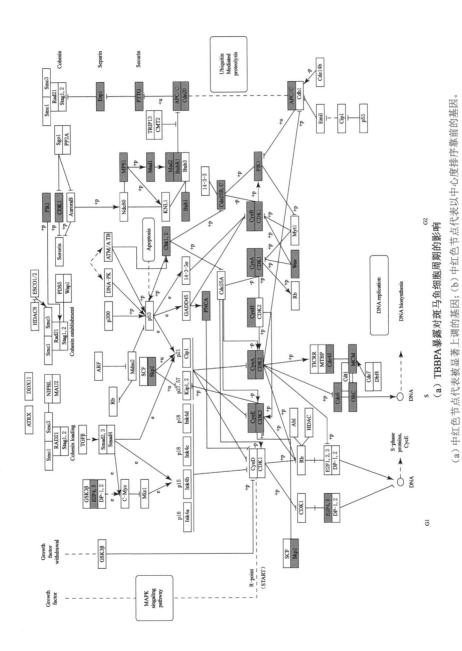

图3-6 TBBPA暴露对斑马鱼细胞周期的影响及富集于细胞周期通路的差异基因的PPI网络分析

(a) TBBPA暴露对斑马鱼细胞周期的影响; (b) 中红色节点代表以中心度排序靠前的基因。

(a) 中红色节点代表显著上调的基因;

第 3 章 聚乙烯微塑料对四溴双酚 A 斑马鱼毒性影响机制

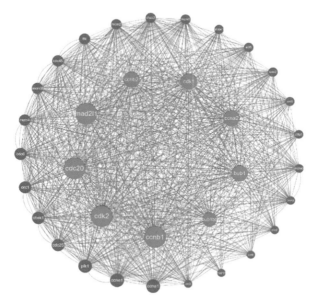

(b) 富集于细胞周期通路的差异表达基因的PPI网络分析

图 3-6 （续）

原因,并可导致脑组织缺陷[20]。细胞周期也是调控斑马鱼心肌增殖的关键因子[21],而编码 Cdks 基因的上调被认为是肿瘤坏死因子-α 诱导主动脉平滑肌细胞增殖过程中的重要机制[22],这表明细胞周期的异常可能最终影响心血管功能。综上所述,TBBPA 诱导斑马鱼的细胞周期紊乱可能会造成多方面的生理影响。

细胞周期的失调最终影响到斑马鱼的 DNA 复制(图 3-7)。斑马鱼细胞中 DNA 复制的起始是通过复制起始处蛋白质复合物的有序组装来实现的,起始的两个步骤为 M 晚期或 G1 早期的起源许可和 S 期的起源激发[图 3-6(a)]。起源识别复合体(origin recognition complex,ORC)作为复制前复合体(pre-RCs)的核心,是一种 6 亚基复合体(ORC1-ORC6),其作用是结合复制起始点[23]。在 M 期和 G1 期,染色质许可和 DNA 复制因子 1(Cdt1)和 CDC6 以 ORC 依赖的方式与复制起点结合,然后协同募集解旋酶微染色体维持复合体(MCM)到起点,完成复制前复合体的组装,结束复制起点许可过程[23]。在复制前复合体组装完成后,起始点获准在 S 期进行复制并准备激发,起始点激发的关键事件是 CDC45 与复制前复合体的结合[24],这一过程通过激酶诱导的磷酸化触发。CDC45 装载到起点后,相关的酶和蛋白质将装载到染色质上启动 DNA 复制[23]。在本研究中,调控 DNA 复制的 *cdcs*、*orcs* 和 *mcms* 显著上调[图 3-6(a)],表明斑马鱼细胞的 DNA 复制过程受到了影响。此外,还检测到作用于斑马鱼有丝分裂 G2 期的检查点激酶(Chks)、cdcs 和 WEE1 样蛋白激酶(WEES)基因的过表达。当 DNA 复制异常发生时,包括 CHK1、WEE1 在内的检查点蛋白激酶将被激活,并触发细胞周期阻滞、复制叉稳定和 DNA 修复机制[25]。上述基因表达上调的最终结果将导致 CDC25s 和 CDKs 的活性受到抑制,从而阻止异常细胞周期的继续[26]。此外,在 G2 期转向 M 期的过程中,检测到有丝分裂纺锤体组装检查点蛋白(*mads*)相关基因的上调表达。作为细胞周期的主要检查点,它们的功能是防止非整倍体的产生和染色体不稳定复制的发生[27]。它们的上调表达将直接影响下游的后期促复合物亚甲基 1(*anapc*1)和 *cdc*20 等参与 M 期的重要基因的表达。而 APC/CDC20 复合物反过来抑制垂体瘤转化基因(*pttg*)的表达,随后影响额外的纺锤体样亚型 X1(*esp*1)基因的表达。PTTG 是染色单体二分过程中必需的蛋白酶,而 ESP1 是分裂黏结蛋白和使姐妹蛋白形成的蛋白酶[28-29]。相关基因的异常表达表明斑马鱼的正常 DNA 复制过程受到了显著干扰。

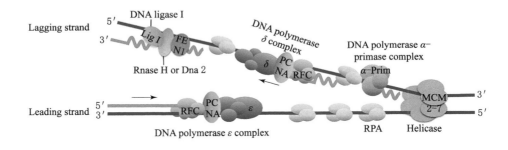

图 3-7 斑马鱼的 DNA 复制过程

3.5 PE-MPs 对 TBBPA 斑马鱼毒性影响的潜在机理

3.5.1 PE-MPs 存在时 TBBPA 诱导的斑马鱼差异表达基因的 GO 富集结果

PE-MPs 与 TBBPA 联合暴露诱导斑马鱼差异表达基因的 GO 富集结果如

图 3-8 所示。统计结果表明，TBBPA+0.25 PE 暴露诱导的斑马鱼差异表达基因显著富集于细胞过程（上调 560，下调 276）、单有机体过程（上调 468，下调 278）、代谢过程（上调 382，下调 160）以及生物调节（上调 300，下调 174）等生物过程，作用于结合（上调 565，下调 284）、催化活性（上调 274，下调 127）、结构分子活性（上调 29，下调 45）、转运活性（上调 37，下调 27）等分子功能，影响细胞（上调 470，下调 215）、细胞部分（上调 470，下调 215）、细胞器（上调 349，下调 130）、膜（上调 249，下调 134）等细胞组分；TBBPA+5 PE 暴露诱导的斑马鱼差异表达基因显著富集于细胞过程（上调 210，下调 103）、单有机体过程（上调 197，下调 99）、生物调节（上调 144，下调 64）以及生物过程调节（上调 137，下调 61）等生物过程，作用于结合（上调 215，下调 91）、催化活性（上调 101，下调 42）、分子转导活性（上调 30，下调 5）、信号转导活性（上调 27，下调 5）等分子功能，影响细胞部分（上调 153，下调 74）、细胞（上调 153，下调 74）、膜（上调 149，下调 46）、膜部分（上调 124，下调 40）等细胞组分；而 TBBPA+20 PE 暴露诱导的斑马鱼差异表达基因显著富集于细胞过程（上调 181，下调 42）、单有机体过程（上调 165，下调 38）、生物调节（上调 115，下调 27）以及生物过程调节（上调 108，下调 25）等生物过程，作用于结合（上调 176，下调 41）、催化活性（上调 94，下调 23）、分子转导活性（上调 29，下调 7）、信号转导活性（上调 27，下调 6）等分子功能，影响细胞部分（上调 139，下调 22）、细胞（上调 139，下调 22）、膜（上调 134，下调 20）、膜部分（上调 117，下调 17）等细胞组分。结果表明，相比于 TBBPA 单独暴露，TBBPA 与 PE-MPs 联合暴露影响的主要细胞组分由细胞器转变为了膜。

TBBPA+0.25 PE 暴露所诱导的差异表达基因在三个 GO 分类中分别富集的显著性最高的十个 GO 条目见表 3-7。富集最显著的生物过程条目与生殖过程相关，包括单生物生殖过程（GO:0044702）、卵子生成（GO:0048477）、雌性配子生成（GO:0007292）等条目；富集最显著的细胞组分条目包括极质（GO:0045495）、非膜结合细胞器（GO:0043228）、细胞内无膜结合细胞器（GO:0043232）等条目；最显著的分子功能条目分别是顶体酶结合（GO:0032190）、细胞骨架的结构组成（GO:0005200）、醇跨膜转运体活性（GO:0015665）等条目。结果表明，相比于 TBBPA 单独暴露，0.25 PE-MPs 的存在改变了主要影响的分子功能条目，但富集的 GO 条目仍表现出显著的遗传毒性特征。

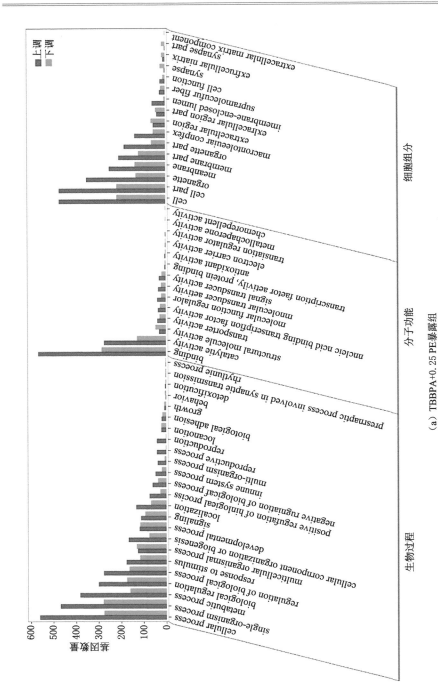

图3-8 PE-MPs与TBBPA联合暴露诱导的斑马鱼差异表达基因的GO二级功能分类统计

(a) TBBPA+0.25 PE暴露组

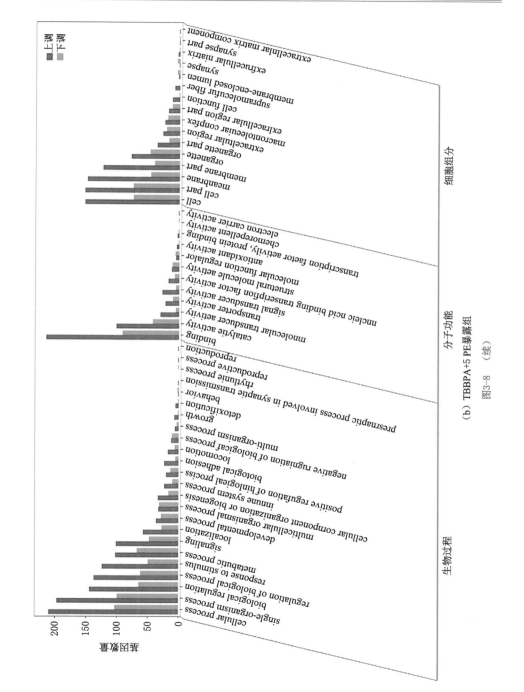

(b) TBBPA+5 PE暴露组

图3-8 （续）

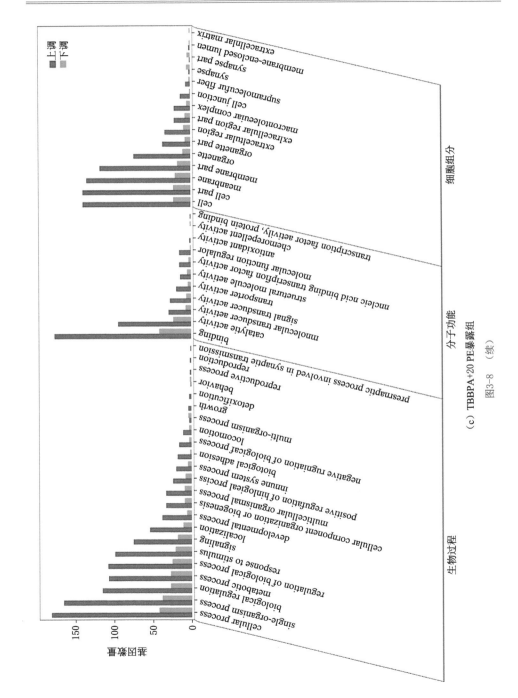

(c) TBBPA+20 PE暴露组

图3-8 （续）

第3章 聚乙烯微塑料对四溴双酚A斑马鱼毒性影响机制

表3-7 TBBPA+0.25 PE暴露诱导差异表达基因的GO富集结果

GO Term	Database	ID	P 值
single organism reproductive process	生物过程	GO:0044702	2.742×10^{-13}
oogenesis	生物过程	GO:0048477	3.122×10^{-13}
female gamete generation	生物过程	GO:0007292	7.596×10^{-13}
reproductive process	生物过程	GO:0022414	8.758×10^{-13}
reproduction	生物过程	GO:0000003	1.159×10^{-12}
cellular process involved in reproduction in multicellular organism	生物过程	GO:0022412	1.243×10^{-12}
germ cell development	生物过程	GO:0007281	6.461×10^{-12}
cell cycle	生物过程	GO:0007049	9.993×10^{-12}
DNA metabolic process	生物过程	GO:0006259	9.764×10^{-11}
gamete generation	生物过程	GO:0007276	1.152×10^{-10}
pole plasm	细胞组分	GO:0045495	0.001 146 789
non-membrane-bounded organelle	细胞组分	GO:0043228	0.089 449 541
intracellular non-membrane-bounded organelle	细胞组分	GO:0043232	0.089 449 541
P granule	细胞组分	GO:0043186	0.001 042 535
germ plasm	细胞组分	GO:0060293	0.001 042 535
microtubule	细胞组分	GO:0005874	0.007 506 255
chromosomal part	细胞组分	GO:0044427	0.018 348 624
microtubule cytoskeleton	细胞组分	GO:0015630	0.018 140 117
chromosome	细胞组分	GO:0005694	0.021 476 23
replication fork	细胞组分	GO:0005657	0.002 293 578
acrosin binding	分子功能	GO:0032190	0.000 834 028
structural constituent of cytoskeleton	分子功能	GO:0005200	0.001 772 31
alcohol transmembrane transporter activity	分子功能	GO:0015665	0.001 146 789
polyol transmembrane transporter activity	分子功能	GO:0015166	0.000 938 282
structural constituent of eye lens	分子功能	GO:0005212	0.001 042 535
glycerol channel activity	分子功能	GO:0015254	0.000 729 775
glycerol transmembrane transporter activity	分子功能	GO:0015168	0.000 729 775
carbohydrate binding	分子功能	GO:0030246	0.007 402 002
nucleocytoplasmic transporter activity	分子功能	GO:0005487	0.000 938 282

TBBPA+5 PE 暴露诱导的差异表达基因在三个 GO 分类中分别富集的显著性最高的十个 GO 条目见表 3-8。富集最显著的生物过程条目分别是超长链脂肪酸的生物合成过程(GO:0042761)、饱和脂肪酸(GO:0019367)、单不饱和脂肪酸(GO:0034625)等脂肪酸相关过程；富集最显著的细胞组分条目分别是膜的组成部分(GO:0016021)、膜的固有成分(GO:0031224)、膜部分(GO:0044425)等膜相关组分；富集最显著的分子功能条目分别是脂肪酸合成酶活性(GO:0004312)、脂肪酸伸长酶活性(GO:0009922)、谷胱甘肽转移酶活性(GO:0004364)等条目。相比于 TBBPA 单独暴露，5 PE-MPs 存在显著改变了主要影响的 GO 条目，主要富集于包括脂肪酸合成与代谢、膜组分和胞间连接等条目。

表 3-8 TBBPA+5 PE 暴露诱导差异表达基因的 GO 富集结果

GO Term	Database	ID	P 值
very long-chain fatty acid biosynthetic process	生物过程	GO:0042761	6.530×10^{-5}
fatty acid elongation, saturated fatty acid	生物过程	GO:0019367	6.530×10^{-5}
fatty acid elongation, monounsaturated fatty acid	生物过程	GO:0034625	6.530×10^{-5}
fatty acid elongation, polyunsaturated fatty acid	生物过程	GO:0034626	6.530×10^{-5}
fatty acid elongation	生物过程	GO:0030497	6.530×10^{-5}
fatty acid elongation, unsaturated fatty acid	生物过程	GO:0019368	6.530×10^{-5}
very long-chain fatty acid metabolic process	生物过程	GO:0000038	9.628×10^{-5}
long-chain fatty-acyl-CoA biosynthetic process	生物过程	GO:0035338	0.000 339 307
fatty-acyl-CoA biosynthetic process	生物过程	GO:0046949	0.000 339 307
long-chain fatty-acyl-CoA metabolic process	生物过程	GO:0035336	0.000 534 138
integral component of membrane	细胞组分	GO:0016021	0.002 171 717
intrinsic component of membrane	细胞组分	GO:0031224	0.002 573 371
membrane part	细胞组分	GO:0044425	0.002 825 334
membrane	细胞组分	GO:0016020	0.003 751 32
intrinsic component of endoplasmic reticulum membrane	细胞组分	GO:0031227	0.004 181 634
cell-cell junction	细胞组分	GO:0005911	0.013 913 89
pigment granule	细胞组分	GO:0048770	0.015 518 6
melanosome	细胞组分	GO:0042470	0.015 518 6
integral component of endoplasmic reticulum membrane	细胞组分	GO:0030176	0.019 110 18
plasma membrane protein complex	细胞组分	GO:0098797	0.028 501 94
fatty acid synthase activity	分子功能	GO:0004312	6.530×10^{-5}

第3章 聚乙烯微塑料对四溴双酚A斑马鱼毒性影响机制

表 3-8(续)

GO Term	Database	ID	P 值
fatty acid elongase activity	分子功能	GO:0009922	6.530×10^{-5}
glutathione transferase activity	分子功能	GO:0004364	0.000 424 73
G-protein coupled receptor binding	分子功能	GO:0001664	0.000 483 376
transferase activity, transferring alkyl or aryl (other than methyl) groups	分子功能	GO:0016765	0.000 596 466
chemokine activity	分子功能	GO:0008009	0.001 968 149
chemokine receptor binding	分子功能	GO:0042379	0.001 968 149
transferase activity, transferring glycosyl groups	分子功能	GO:0016757	0.006 413 503
NAD+ ADP-ribosyltransferase activity	分子功能	GO:0003950	0.006 621 684
calcium activated cation channel activity	分子功能	GO:0005227	0.006 752 257

TBBPA+20 PE暴露诱导的差异表达基因在三个GO分类中分别富集的显著性最高的十个GO条目见表3-9。富集最显著的生物过程条目分别是免疫反应(GO:0006955)、抗原处理和呈递(GO:0019882)、有机酸生物合成过程(GO:0016053)等免疫相关条目；富集最显著的细胞组分条目分别是膜整体组分(GO:0016021)、细胞-细胞连接(GO:0005911)、膜的固有成分(GO:0031224)等细胞结构与膜相关条目；富集最显著的分子功能条目分别是蛋白激酶C活性(GO:0004697)、丝氨酸型内肽酶活性(GO:0004252)、肾上腺素受体活性(GO:0004935)等条目。

表 3-9 TBBPA+20 PE暴露诱导差异表达基因的GO富集结果

GO Term	Database	ID	P 值
immune response	生物过程	GO:0006955	0.000 423 079
antigen processing and presentation	生物过程	GO:0019882	0.000 470 727
organic acid biosynthetic process	生物过程	GO:0016053	0.004 908 657
carboxylic acid biosynthetic process	生物过程	GO:0046394	0.004 908 657
adenylate cyclase-activating adrenergic receptor signaling pathway	生物过程	GO:0071880	0.013 087 82
adrenergic receptor signaling pathway	生物过程	GO:0071875	0.013 087 82
regulation of heart contraction	生物过程	GO:0008016	0.017 625 47
actin polymerization or depolymerization	生物过程	GO:0008154	0.017 788 09

表 3-9(续)

GO Term	Database	ID	P 值
vascular smooth muscle cell development	生物过程	GO:0097084	0.017 931 61
chloride ion homeostasis	生物过程	GO:0055064	0.017 931 61
integral component of membrane	细胞组分	GO:0016021	2.880×10^{-5}
cell-cell junction	细胞组分	GO:0005911	3.160×10^{-5}
intrinsic component of membrane	细胞组分	GO:0031224	3.552×10^{-5}
bicellular tight junction	细胞组分	GO:0005923	4.465×10^{-5}
occluding junction	细胞组分	GO:0070160	5.225×10^{-5}
apical junction complex	细胞组分	GO:0043296	5.225×10^{-5}
membrane part	细胞组分	GO:0044425	7.691×10^{-5}
membrane	细胞组分	GO:0016020	0.000 314 975
cell junction	细胞组分	GO:0030054	0.001 468 111
MHC class Ⅱ protein complex	细胞组分	GO:0042613	0.008 337 625
protein kinase C activity	分子功能	GO:0004697	0.005 160 574
serine-type endopeptidase activity	分子功能	GO:0004252	0.009 302 593
adrenergic receptor activity	分子功能	GO:0004935	0.010 594 2
serine hydrolase activity	分子功能	GO:0017171	0.017 307 92
serine-type peptidase activity	分子功能	GO:0008236	0.017 307 92
cysteinyl leukotriene receptor activity	分子功能	GO:0001631	0.017 931 61
argininosuccinate synthase activity	分子功能	GO:0004055	0.017 931 61
amidinotransferase activity	分子功能	GO:0015067	0.017 931 61
glycine amidinotransferase activity	分子功能	GO:0015068	0.017 931 61
beta2-adrenergic receptor activity	分子功能	GO:0004941	0.017 931 61

相比于 TBBPA 暴露组的 GO 功能网络,PE-MPs 的存在显著影响了 GO 功能网络的特征(图 3-9)。具体而言,PE-MPs 的加入以浓度相关性的方式降低了 TBBPA 诱导的差异表达基因形成的 GO 功能分组网络的节点数和复杂度。在网络功能方面,TBBPA+0.25 PE 存在对网络功能的影响较小,细胞周期进程、DNA 复制和卵子等也是 TBBPA+0.25 PE 暴露组的主要功能网络;TBBPA+5 PE 暴露组的网络主要与 G 蛋白偶联受体结合有关。G 蛋白偶联受体作为重要的跨膜结构域蛋白,通过细胞内级联反应发挥重要的生理功能,参与细胞色素 P450 介导的解毒和免疫细胞活化等免疫功能的调节[30-31]。TBBPA+20 PE 组的网络功能与机体炎症反应相关,如质膜的外源性成分、高尔基体膜、对细胞因子的反应等。GO 网络分析的结果表明,PE-MPs 的存在呈浓度依赖性地降低了 TBBPA 的遗传毒性影响,而加重了免疫毒性影响。

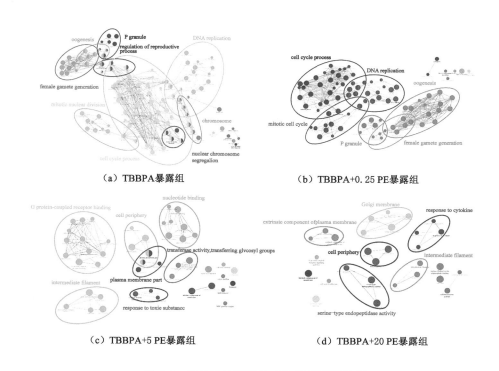

图 3-9 差异表达基因的 GO 富集网络分析

3.5.2 PE-MPs 存在时 TBBPA 诱导的斑马鱼差异表达基因的通路富集结果

联合暴露组的差异表达基因的通路富集分析结果如图 3-10 所示。TBBPA+0.25 PE 暴露组差异表达基因富集的通路功能与 TBBPA 暴露组相似,主要与细胞周期与遗传相关,如细胞周期通路、DNA 复制通路、孕酮介导的卵母细胞成熟通路等。而随着 PE-MPs 浓度增加,联合暴露组的差异表达基因富集的通路出现了明显的差异。TBBPA+5 PE 暴露组差异表达基因富集的通路为活性氧的解毒、药物代谢-细胞色素 P450 及谷胱甘肽代谢等;TBBPA+20 PE 暴露组差异表达基因富集的通路包括 PIP2 的水解效应、白细胞介素-17 信号以及炎症小体等通路。相关结果表明,低浓度 PE-MPs 存在并未显著改变 TBBPA 暴露影响的斑马鱼的相关功能,而随着 PE-MPs 浓度的提高,TBBPA 暴露诱导的斑马鱼差异表达基因的功能开始发生明显变化,这表明 PE-MPs 的存在会显著影响 TBBPA 对斑马鱼基因层面的毒性效应。

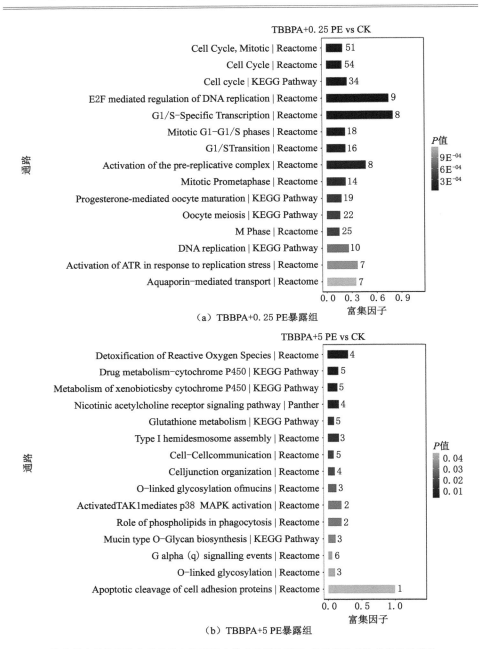

(a) TBBPA+0.25 PE暴露组

(b) TBBPA+5 PE暴露组

横坐标表示注释的差异基因占该通路中所有基因的频率,纵坐标为各数据库以显著性排序最显著富集的15个通路;数字代表富集基因的数量;颜色代表显著性。

图 3-10　PE-MPs 与 TBBPA 联合暴露组诱导差异表达基因的通路富集分析

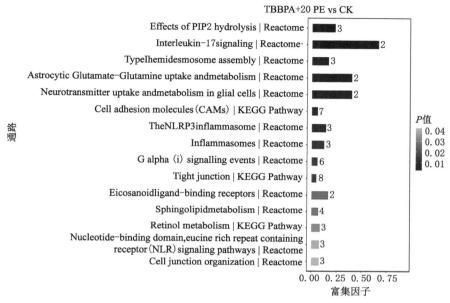

(c) TBBPA+20 PE暴露组

图 3-10 （续）

对不同暴露组富集于 KEGG 数据库的差异表达基因的二级功能注释表明，PE-MPs 的加入显著改变了富集于不同二级功能的差异表达基因数量[图 3-11(a)]。对于各暴露组所有显著富集（$P<0.05$）的 KEGG 通路进行统计，结果如图 3-11(b)所示。结果表明，高浓度 PE-MPs 的加入显著降低了差异表达基因富集通路的数量。在功能方面，TBBPA＋0.25 PE 暴露组与 TBBPA 暴露组相似，也表现出显著的遗传毒性影响特征。而随着共暴露 PE-MPs 浓度的增加，联合暴露组的差异表达基因不再富集于遗传与细胞周期相关通路，而是表现出了对免疫系统的影响。TBBPA＋5 PE 暴露组显著诱导了细胞色素 P450 相关通路和谷胱甘肽代谢通路的调节。细胞色素 P450 参与内源性物质和包括药物、环境化合物在内的外源性物质的代谢环境毒素的代谢，其表达和活性受免疫反应的影响[32]，而谷胱甘肽参与细胞免疫调节，保护细胞内大分子免受氧化损伤[33]。TBBPA＋20 PE 组中显著富集的通路是细胞黏附分子和紧密连接通路，它们是生物体炎症反应中影响白细胞迁移的重要因素[34]。总而言之，通路分析结果表明随着共暴露 PE-MPs 浓度的增加，TBBPA 的遗传毒性得到减轻，但免疫毒性被显著加重，这印证了前一章中生化指标测定的结果。

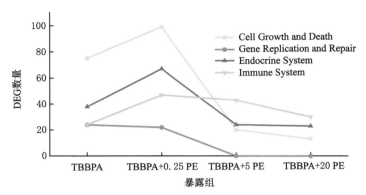

(a) 不同暴露组差异表达基因KEGG二级功能分类的数量差异

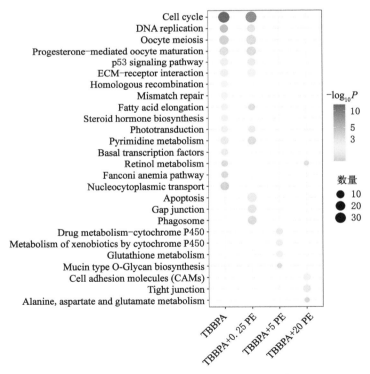

(b) 不同暴露组差异表达基因富集的KEGG通路

点大小代表基因数量；颜色代表显著性。

图 3-11　PE-MPs 对 TBBPA 暴露诱导差异表达基因 KEGG 通路富集结果的影响

3.5.3 PE-MPs 对 TBBPA 诱导的斑马鱼细胞周期与 DNA 复制通路异常的影响

PE-MPs 对 TBBPA 诱导细胞周期通路异常的影响如图 3-12 所示。

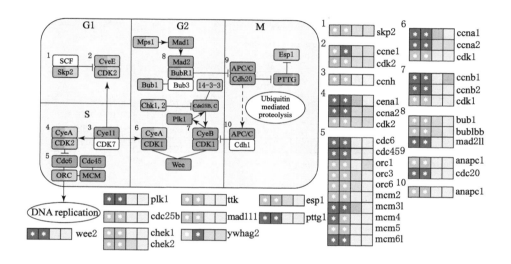

热图由左至右分别代表 TBBPA 暴露组、TBBPA+0.25 PE 暴露组、TBBPA+5 PE 暴露组、TBBPA+20 PE 暴露组;红色代表基因上调,蓝色代表基因下调,颜色深度代表差异表达程度;"＊"标记代表显著差异表达的基因。

图 3-12 PE-MPs 对 TBBPA 诱导细胞周期通路异常的影响

相比于 TBBPA 单独暴露,TBBPA+0.25 PE 暴露组的差异基因表达模式与 TBBPA 单独暴露组相似,这表明低浓度 PE-MPs 加入对 TBBPA 诱导细胞周期通路异常的影响较为轻微。而随着 PE-MPs 浓度增加,TBBPA 诱导细胞周期通路异常的影响受到了显著影响,具体表现为基因差异表达水平及显著性的明显下降,这表明高 PE-MPs 的加入显著降低了 TBBPA 诱导的斑马鱼细胞周期异常。这一效应在 PE-MPs 加入对 TBBPA 诱导 DNA 复制通路异常的影响上得也到了体现(图 3-13)。

在 DNA 复制通路中,TBBPA+0.25 PE 暴露组的差异基因也具有 TBBPA 单独暴露组相似的表达模式,而随着 PE-MPs 浓度增加,TBBPA 诱导 DNA 复制通路的基因差异也在表达水平及显著性上出现了明显的下降。综上所述,

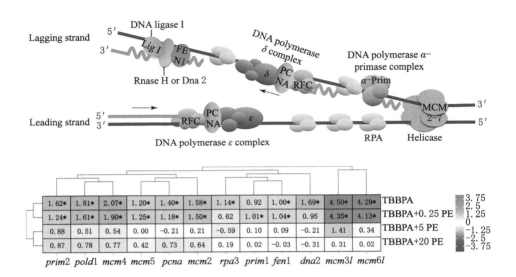

热图红色代表基因上调,蓝色代表基因下调,颜色深度代表差异表达程度;
"*"标记代表显著差异表达的基因。

图 3-13 PE-MPs 对 TBBPA 诱导 DNA 复制通路异常的影响

TBBPA 的遗传毒性随着 PE-MPs 共暴露浓度的增加而减轻。其机理可能是 PE-MPs 共暴露减少了 TBBPA 的积累。研究表明,肠道中清洁的 MPs 对有机化学物质具有较高的吸附亲和力,有利于有机污染物的转移,从而减轻机体的解毒负担,这就是 MPs 的"清洁效应"[35]。考虑到 PE-MPs 与 TBBPA 的相互作用,PE-MPs 可能通过吸附斑马鱼肠道中游离的 TBBPA 来降低 TBBPA 的破坏作用,从而降低 TBBPA 的遗传毒性。Yang 等[36]的研究表明,PS-MPs 可以通过吸附减少氯化聚氟烷基醚磺酸盐(F-53B)的生物蓄积。Ziajahromi 等[37]的研究也表明 PE-MPs 可以通过降低联苯菊酯的生物利用度来降低其对蠓幼虫的毒性。这为本研究的结论提供了依据。

3.5.4　PE-MPs 对斑马鱼免疫相关通路的影响

与对遗传毒性的影响不同,PE-MPs 显著诱导了差异表达基因在多条免疫相关通路的富集(表 3-10)。

第 3 章 聚乙烯微塑料对四溴双酚 A 斑马鱼毒性影响机制

表 3-10 复合暴露对斑马鱼免疫通路的激活

	TBBPA+0.25 PE	TBBPA+5 PE	TBBPA+20 PE
KEGG Pathway	Phagosome	Glutathione metabolism Drug metabolism cytochrome P450 Metabolism of xenobiotics by cytochrome P450	Cell adhesion molecules(CAMs) Tight junction
Reactome	—	Detoxification of reactive oxygen species	Interleukin-17 signaling The NLRP3 inflammasome

具体而言,TBBPA+0.25 PE 暴露诱导了差异表达基因在吞噬体通路的显著富集。吞噬体是在胞吞作用中在被吞噬物质周围形成的囊泡,是一种在免疫过程中常见的细胞结构,其与活性氧(ROS)的过量产生和吞噬细胞的免疫应答密切相关,是生物体有效抵御感染的重要标志[38]。TBBPA+5 PE 暴露诱导了差异表达基因在细胞色素 P450、谷胱甘肽代谢及活性氧的解毒通路的显著富集。细胞色素 P450 酶介导的 TBBPA 生物转化是斑马鱼中 TBBPA 解毒的重要机制[39-40],而谷胱甘肽是维持免疫功能和清除 ROS 的重要调节因子[41]。斑马鱼细胞内的活性氧的解毒过程则首先由超氧化物歧化酶(SOD)催化产生 H_2O_2,H_2O_2 进一步被过氧化氢酶(CAT)、谷胱甘肽(GSH)、硫氧还蛋白(TXN)等降解为 H_2O 和 O_2,其激活表明机体受到外界刺激导致过量 ROS 的产生(图 3-14),相关通路的激活表明了 TBBPA+5 PE 诱导了机体的免疫反应。

TBBPA+20 PE 暴露诱导了细胞黏附分子(CAMs)、紧密连接、白细胞介素-17 信号和 NLRP3 炎症小体通路的激活。炎症小体的组装响应于多种因素,包括细胞结构的破坏和细胞循环的失衡(图 3-15)。白细胞介素-17 是免疫细胞分泌的细胞因子,在宿主对黏膜感染的防御反应中具有重要作用[42],而 CAMs 和紧密连接是影响免疫反应中白细胞跨内皮迁移的重要因素[43]。相关通路的激活表明 TBBPA+20 PE 暴露诱发了炎症等免疫反应,导致了更严重的免疫毒性。

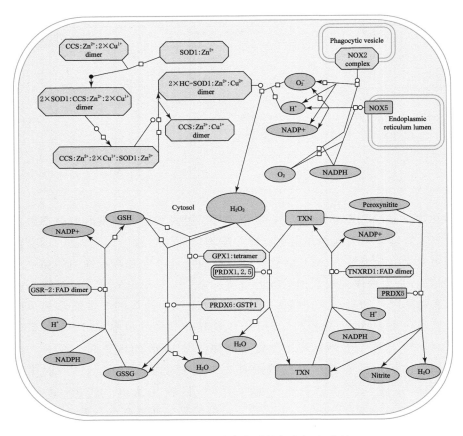

图 3-14 斑马鱼细胞内的活性氧的解毒过程

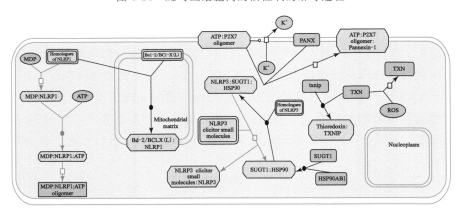

图 3-15 斑马鱼炎症小体的形成

3.6 转录组测序结果验证

采用 RT-qPCR 实验验证转录组测序结果的可靠性。转录组测序结果表明，TBBPA 暴露组和 TBBPA+0.25 PE 暴露组之间存在大量共享的差异表达基因，因此采用将 TBBPA 组和 TBBPA+0.25 PE 组、TBBPA+5 PE 组和 TBBPA+20 PE 组分别选取基因进行分组验证的方法。选取 TBBPA 组和 TBBPA+0.25 PE 组的 4 个差异表达基因（*prim*2、*thrb*、*itga*11a 和 *cyp*17a1）以及 TBBPA+5 PE 组和 TBBPA+20 PE 组的 4 个差异表达基因（*prdx*1、*nphs*2、*klf*3 和 *tmprss*15）进行 RT-qPCR 实验。这些基因为调控斑马鱼细胞分裂、遗传、免疫及抗氧化等功能的典型基因，结果如图 3-16 和图 3-17 所示。结果表明，RT-qPCR 测定的基因差异表达趋势与转录组测序结果相一致。相关性分析结果表明，两种方法测定的基因相对表达量具有一定的相关性（$R^2=0.6379$），这验证了转录组测序数据的可靠性。

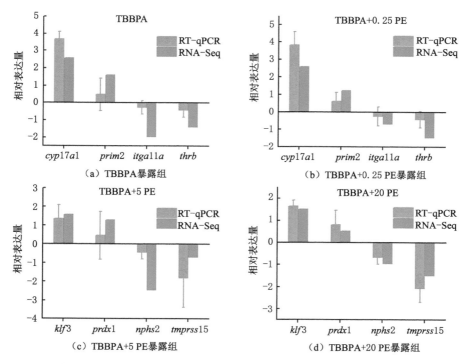

纵坐标为基因的相对表达量；RT-qPCR 结果以平均值±标准差表示（$n=3$）。

图 3-16 RT-qPCR 结果与转录组测序结果对比

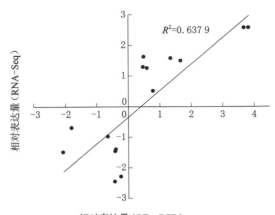

图 3-17 RT-qPCR 结果与转录组测序结果的相关性

3.7 TBBPA 对斑马鱼内分泌干扰效应验证

3.7.1 TBBPA 对斑马鱼甲状腺激素受体 β(THRβ) 的干扰效应

THRβ 是一种核受体,其能通过调节下游细胞周期蛋白的表达来对细胞周期进行调控。先前的体外研究已经证实 TBBPA 可能具有对 THRβ 的正常功能的干扰作用(图 3-18)[44]。在本研究中,TBBPA 诱导了编码 THRβ 的基因 thrb 的显著下调(FC 值为 0.28,$P<0.05$),表明 TBBPA 可能通过干扰 THRβ 诱导细胞周期失调。针对 THRβ 与其原配体 T_3 及 T_4 的竞争对接与分子动力学模拟结果见表 3-11 及图 3-19。THRβ 的结合腔半径及体积分别为 10.90 Å 与 278.75 Å^3。相互作用分析表明,TBBPA 主要通过疏水作用与 THRβ 结合,同时 THRβ 的 HIS369 和 MET247 残基与 TBBPA 形成氢键。T_3 和 T_4 与 THRβ 的结合构象的氢键数多于 TBBPA,疏水相互作用数少于 TBBPA。分子动力学模拟结果表明,在与 THRβ 的相互作用方面,TBBPA-THRβ 的 30 ns RMSD 值高于 T_3-THRβ 和 T_4-THRβ,表明 T_3 和 T_4 可能与 THRβ 有比 TBBPA 更有效的结合构象。T_3-THRβ 的 RMSD 值达到稳定状态的时间最长,说明与 TBBPA 和 T_4 相比,T_3 和 THRβ 的结合状态不稳定。三种配体与 THRβ 的结合自由能排序为 T_4(−191.797 kJ/mol)<TBBPA(−179.531 kJ/mol)<T_3(−171.423 kJ/mol)。总而言之,分子对接与分子动力学模拟结果证实了 TBBPA 可能通过干扰 THRβ 影响斑马鱼的细胞周期与 DNA 复制。

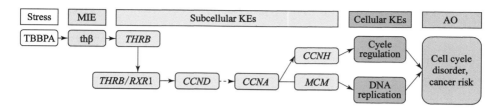

图 3-18 TBBPA 通过干扰 THRβ 影响细胞周期与 DNA 复制的不良结局途径

表 3-11 TBBPA 与 THRβ 的对接参数与结果

蛋白	结合腔体积 /Å³	结合腔半径 /Å	配体	作用力			结合自由能 /(kJ/mol)
				氢键	疏水相互作用	卤素相互作用	
THRβ	278.75	10.90	TBBPA	2	20	1	−179.531
			T_3	4	11	1	−171.423
			T_4	6	13	2	−191.797

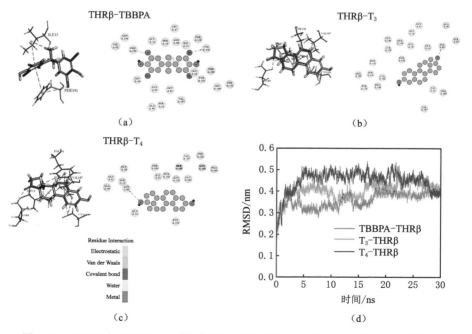

图 3-19 THRβ 与 TBBPA 及其配体的分子对接构象与 30 ns 分子动力学模拟结果

3.7.2　TBBPA 对斑马鱼 17β-羟基类固醇脱氢酶 1(HSD17β1)的干扰效应

HSD17β1 是一种重要的雌激素合成酶,参与类固醇激素的生物合成过程,是调控卵母细胞成熟与减数分裂的重要上游蛋白[45]。针对 HSD17β1 与其原配体 E_1 及 E_2 的竞争对接与分子动力学模拟结果见表 3-12 及图 3-20。

表 3-12　TBBPA 与 HSD17β1 的对接参数与结果

蛋白	结合腔体积 /Å³	结合腔半径 /Å	配体	作用力			结合自由能 /(kJ/mol)
				氢键	疏水相互作用	卤素相互作用	
HSD17β1	802.625	17.40	TBBPA	—	4	—	−130.843
			E_1	1	8	—	−90.309
			E_2	1	9	—	−96.861

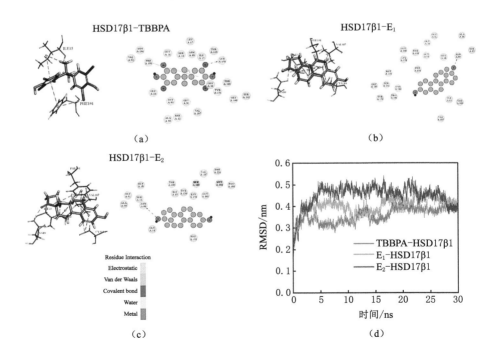

图 3-20　HSD17β1 与 TBBPA 及其配体的分子对接构象与 30 ns 分子动力学模拟结果

HSD17β1 的结合腔半径及体积分别为 17.40 Å 与 802.625 Å³。相互作用分析表明,TBBPA 与 HSD17β1 的结合构象没有形成氢键,主要的作用方式是疏水作用且相互作用的残基数少于 HSD17β1 与 E_1 和 E_2 的结合构象。分子动力学模拟结果表明,在与 HSD17β1 的相互作用方面,TBBPA-HSD17β1、E_1-HSD17β1 和 E_2-HSD17β1 的 30 ns RMSD 值均接近 0.4 nm,这表明 TBBPA 对雌激素合成可能具有干扰作用。三种配体与 THRβ 的结合自由能排序为 TBBPA(−130.843 kJ/mol)<E_2(−96.861 kJ/mol)<E_1(−90.309 kJ/mol)。结果表明,TBBPA 可能通过其与雌激素及其前体结构的类似性影响斑马鱼雌激素的生物合成,进而对卵母细胞成熟与减数分裂产生间接影响。

3.8 本章小结

本章采用转录组测序技术与生物信息学方法,研究了 TBBPA 对斑马鱼的毒性机制及 PE-MPs 对 TBBPA 毒性的影响。通过实时荧光定量 PCR、分子对接和分子动力学模拟等方法,对相关结果进行了进一步验证。主要结论如下:

① 相比于对照组,TBBPA 暴露组诱导了 1 341 个基因显著差异表达,其中上调 941 个、下调 400 个;TBBPA+0.25 PE 暴露组诱导了 1 835 个基因显著差异表达,其中上调 1 201 个、下调 634 个;TBBPA+5 PE 暴露组诱导了 758 个基因显著差异表达,其中上调 507 个、下调 251 个;TBBPA+20 PE 暴露组诱导了 554 个基因显著差异表达,其中上调 429 个、下调 125 个,PE-MPs 的加入显著改变了 TBBPA 诱导的差异表达基因数量。

② TBBPA 暴露组的生物信息学分析表明,TBBPA 暴露诱导的差异表达基因在细胞周期、DNA 复制及卵母细胞减数分裂等通路富集,具有显著的遗传毒性效应。GSEA 分析验证了 TBBPA 对相关通路的上调作用,而通路网络分析则表明 TBBPA 可能通过内分泌干扰途径介导斑马鱼的遗传毒性产生。PPI 分析表明,CCNB、CDK2、CDC20、MAD2L1 等蛋白可能在斑马鱼的细胞周期紊乱中起关键作用。

③ 复合暴露组的生物信息学分析表明,PE-MPs 的加入显著缓解了 TBBPA 诱导的遗传毒性,具体表现在随着共暴露 PE-MPs 浓度的增加,复合暴露组诱导的差异表达基因不再富集于遗传毒性相关的通路。但 PE-MPs 的加入显著诱导免疫相关通路的激活,包括 TBBPA+0.25 PE 暴露诱导的吞噬体通路、TBBPA+5 PE 暴露诱导的细胞色素 P450、谷胱甘肽代谢及活性氧的解毒通路以及 TBBPA+20 PE 暴露诱导的细胞黏附分子(CAMs)、紧密连接、白细胞介素-17 信号和 NLRP3 炎症小体通路的激活,这表明 PE-MPs 的加入导致了

TBBPA 的免疫毒性加重。

④ RT-qPCR 实验结果表明，选定的 8 个基因（$prim2$、$thrb$、$itga11a$、$cyp17a1$、$prdx1$、$nphs2$、$klf3$ 和 $tmprss15$）在相应暴露组中的差异表达趋势与转录组测序结果相同，且存在一定相关性（$R^2=0.6379$），这验证了转录组测序数据的可靠性。

⑤ 分子对接与分子动力学模拟结果表明，TBBPA 与 THRβ 和 HSD17β1 均能形成稳定的结合构象，并对这两种蛋白的原配体产生干扰作用，这表明 TBBPA 能通过干扰核受体影响斑马鱼细胞周期及 DNA 复制的正常进行，以及通过干扰雌激素合成影响斑马鱼卵母细胞的成熟与减数分裂。

参考文献

[1] KANEHISA M,GOTO S,SATO Y,et al.KEGG for integration and interpretation of large-scale molecular data sets[J].Nucleic acids research,2012,40(1):109-114.

[2] CROFT D,O'KELLY G,WU G M,et al.Reactome:a database of reactions,pathways and biological processes[J].Nucleic acids research,2011,39(1):691-697.

[3] MI H Y,MURUGANUJAN A,THOMAS P D.PANTHER in 2013:modeling the evolution of gene function,and other gene attributes,in the context of phylogenetic trees[J].Nucleic acids research,2013,41(1):377-386.

[4] MA Z W,WU Y L,JIN J L,et al.Phylogenetic analysis reveals the evolution and diversification of cyclins in eukaryotes[J].Molecular phylogenetics and evolution,2013,66(3):1002-1010.

[5] PAVLETICH N P.Mechanisms of cyclin-dependent kinase regulation:structures of Cdks,their cyclin activators,and Cip and INK4 inhibitors[J].Journal of molecular biology,1999,287(5):821-828.

[6] ASGHAR U,WITKIEWICZ A K,TURNER N C,et al.The history and future of targeting cyclin-dependent kinases in cancer therapy[J].Nature reviews drug discovery,2015,14:130-146.

[7] MORGAN D O.Cyclin-dependent kinases:engines,clocks,and microprocessors[J].Annual review of cell and developmental biology,1997,13:261-291.

[8] DHAVAN R,TSAI L H.A decade of CDK5[J].Nature reviews molecular cell biology,2001,2:749-759.

[9] BLOOM J,CROSS F R.Multiple levels of cyclin specificity in cell-cycle

control[J].Nature reviews molecular cell biology,2007,8:149-160.
[10] CAMINS A,PIZARRO J G,FOLCH J.Cyclin-dependent kinases[M]//Brenner's Encyclopedia of Genetics.Amsterdam:Elsevier,2013:260-266.
[11] MALUMBRES M,BARBACID M.Cell cycle,CDKs and cancer:a changing paradigm[J].Nature reviews cancer,2009,9:153-166.
[12] SAUSVILLE E A.Complexities in the development of cyclin-dependent kinase inhibitor drugs[J].Trends in molecular medicine,2002,8(4):32-37.
[13] BALCZON R.Overexpression of cyclin A in human HeLa cells induces detachment of kinetochores and spindle pole/centrosome overproduction[J].Chromosoma,2001,110(6):381-392.
[14] TANE S,CHIBAZAKURA T.Cyclin A overexpression induces chromosomal double-strand breaks in mammalian cells[J]. Cell cycle, 2009, 8(23): 3900-3903.
[15] KANG J S,BONG J,CHOI J S,et al.Differentially transcriptional regulation on cell cycle pathway by silver nanoparticles from ionic silver in larval zebrafish (*Danio rerio*)[J].Biochemical and biophysical research communications,2016,479(4):753-758.
[16] LU L R,HU J J,LI G Y,et al.Low concentration Tetrabromobisphenol A (TBBPA) elevating overall metabolism by inducing activation of the Ras signaling pathway[J].Journal of hazardous materials,2021,416:125797.
[17] MATTHEWS H K,BERTOLI C,DE BRUIN R A M.Cell cycle control in cancer[J].Nature reviews molecular cell biology,2022,23:74-88.
[18] KATO T,DAIGO Y,ARAGAKI M,et al.Overexpression of CDC20 predicts poor prognosis in primary non-small cell lung cancer patients[J].Journal of surgical oncology,2012,106(4):423-430.
[19] WANG W,BU B T,XIE M J,et al.Neural cell cycle dysregulation and central nervous system diseases[J].Progress in neurobiology,2009,89(1):1-17.
[20] YANG Y,HERRUP K.Cell division in the CNS:protective response or lethal event in post-mitotic neurons? [J].Biochimica et biophysica acta,2007,1772(4):457-466.
[21] ZUPPO D A,TSANG M.Zebrafish heart regeneration:factors that stimulate cardiomyocyte proliferation[J].Seminars in cell and developmental biology,2020,100:3-10.
[22] FAN S L,LI X,LIN J,et al.Honokiol inhibits tumor necrosis factor-α-

stimulated rat aortic smooth muscle cell proliferation via caspase-and mitochondrial-dependent apoptosis[J].Inflammation,2014,37(1):17-26.

[23] BELL S P,DUTTA A.DNA replication in eukaryotic cells[J].Annual review of biochemistry,2002,71:333-374.

[24] MACHIDA Y J,DUTTA A.Cellular checkpoint mechanisms monitoring proper initiation of DNA replication[J].Journal of biological chemistry,2005,280(8):6253-6256.

[25] GRALEWSKA P,GAJEK A,MARCZAK A,et al.Participation of the ATR/CHK1 pathway in replicative stress targeted therapy of high-grade ovarian cancer[J].Journal of hematology and oncology,2020,13(1):39.

[26] CLEARY J M,AGUIRRE A J,SHAPIRO G I,et al.Biomarker-guided development of DNA repair inhibitors[J].Molecular cell,2020,78(6):1070-1085.

[27] RYAN S D,BRITIGAN E M C,ZASADIL L M,et al.Up-regulation of the mitotic checkpoint component Mad1 causes chromosomal instability and resistance to microtubule poisons[J].Proceedings of the national academy of sciences of the United States of America,2012,109(33):2205-2214.

[28] HO K L,MA L N,CHEUNG S,et al.A role for the budding yeast separase,Esp1, in Ty1 element retrotransposition [J]. PLoS genetics, 2015, 11(3):1005109.

[29] MUSACCHIO A,SALMON E D.The spindle-assembly checkpoint in space and time[J].Nature reviews molecular cell biology,2007,8:379-393.

[30] LI T,LIU N N.The function of G-protein-coupled receptor-regulatory cascade in southern house mosquitoes (*Diptera*:*Culicidae*)[J].Journal of medical entomology,2018,55(4):862-870.

[31] MAYADAS T N,CULLERE X,LOWELL C A.The multifaceted functions of neutrophils[J].Annual review of pathology,2014,9:181-218.

[32] WANG G Y,XIAO B,DENG J Y,et al.The role of cytochrome P450 enzymes in COVID-19 pathogenesis and therapy[J].Frontiers in pharmacology,2022,13:791922.

[33] MORRIS G,GEVEZOVA M,SARAFIAN V,et al.Redox regulation of the immune response[J].Cellular and molecular immunology,2022,19:1079-1101.

[34] MULLER W A.Leukocyte-endothelial-cell interactions in leukocyte transmigration and the inflammatory response[J]. Trends in immunology,

2003,24(6):327-334.

[35] SCHELL T,RICO A,CHERTA L,et al.Influence of microplastics on the bioconcentration of organic contaminants in fish:is the "Trojan horse" effect a matter of concern? [J].Environmental pollution,2022,306:119473.

[36] YANG H L,LAI H,HUANG J,et al.Polystyrene microplastics decrease F-53B bioaccumulation but induce inflammatory stress in larval zebrafish [J].Chemosphere,2020,255:127040.

[37] ZIAJAHROMI S,KUMAR A,NEALE P A,et al.Effects of polyethylene microplastics on the acute toxicity of a synthetic pyrethroid to midge larvae (*Chironomus tepperi*) in synthetic and river water[J].Science of the total environment,2019,671:971-975.

[38] NÜSSE O.Biochemistry of the phagosome:the challenge to study a transient organelle[J].The scientific world journal,2011,11:2364-2381.

[39] SHEN M N,CHENG J,WU R H,et al.Metabolism of polybrominated diphenyl ethers and tetrabromobisphenol A by fish liver subcellular fractions *in vitro*[J].Aquatic toxicology,2012,114/115:73-79.

[40] 陈秋霞,曾苏.斑马鱼在药物代谢中的研究进展[J].药学学报,2011,46(9):1026-1031.

[41] MASSARSKY A,KOZAL J S,DI GIULIO R T.Glutathione and zebrafish:old assays to address a current issue[J].Chemosphere,2017,168:707-715.

[42] HUANGFU L J,LI R Y,HUANG Y M,et al.The IL-17 family in diseases:from bench to bedside[J].Signal transduction and targeted therapy,2023,8:402.

[43] ENGELHARDT B,WOLBURG H.Mini-review:Transendothelial migration of leukocytes:through the front door or around the side of the house? [J].European journal of immunology,2004,34(11):2955-2963.

[44] WANG X Q,LI F,CHEN J W,et al.Integration of computational toxicology, toxicogenomics data mining,and omics techniques to unveil toxicity pathways [J].ACS sustainable chemistry and engineering,2021,9(11):4130-4138.

[45] CHEN S L,WANG S W,ZHENG J Y,et al.Bisphenol analogues inhibit human and rat 17β-hydroxysteroid dehydrogenase 1:3D-quantitative structure-activity relationship (3D-QSAR) and *in silico* docking analysis[J].Food and chemical toxicology:an international journal published for the british industrial biological research association,2023,181:114052.

第4章 微塑料对全氟辛酸的毒性影响综述

4.1 选题背景

近70年来,随着多氟化合物的广泛应用,全氟烷基和多氟烷基物质(PFAS)因其出色的热稳定性、化学稳定性和疏水疏油性等特性而被广泛用于航空航天、国防、汽车、服装、建筑和医疗等领域。目前,已经在水体[1]、土壤[2]、沉积物[3]和空气[4]等多种环境中检测到PFAS的存在。而全氟辛酸(PFOA)和全氟辛烷磺酸(PFOS)作为环境中最常被检测到的PFAS[5],因其高稳定性、高流动性和可生物蓄积性等特性,已被《斯德哥尔摩公约》列为持久性有机污染物而备受关注[6-8]。早在20世纪90年代,PFOA就已经在环境中被检测到,之后人们对其研究也从未停止。氟化工厂作为环境中PFOA的重要来源,在靠近工厂的地表水、地下水、土壤、植被、饮用水和食物样本中,均检测到了较高浓度的PFOA,且最高浓度超过了PFOA的标准和限值(土壤中的PFOA除外)[9]。Quiñones等[10]对从美国多地收集的饮用水处理设施样本以及相关地表、地面和废水源的中的多氟化合物含量测定结果显示,PFOA的总浓度在任何地点都是最高的。PFOA还存在于食物[11]、饮用水[12]以及生物体中[13-14],这意味着PFOA可以进入人体并造成健康风险,目前已经从世界许多地区的人体中检测到PFOA的存在[15]。Gebbink等[9]对氟化工厂附近居民和工厂工作人员血液中的PFOA含量测定发现,部分人员血液中PFOA超过安全水平。对于饮用水中能检测到PFOA的地区,Gyllenhammar等[16]在该地区儿童体内检测到较高浓度的PFOA。PFOA的环境存留量及其带来的健康威胁使得人们开始限制其使用。美国环境署(EPA)建议PFOA、PFOS或其混合物浓度不应超过70 ng/L[17-18],此外,美国9个州分别制定了更加严格的PFOA指南[19]。21世纪初,明尼苏达矿业及机器制造公司和杜邦等公司逐渐停止使用PFOA,尽管如此,环境中仍能检测到高浓度的PFOA[20-22]。因此,PFOA对环境和生物的影响应得到进一步

第 4 章 微塑料对全氟辛酸的毒性影响综述

重视。

近几十年来,塑料制品被广泛应用于工业、农业、建筑、包装和电子器材等各类生产生活领域,但其广泛的应用也带来了大量的废弃物。截止到 2015 年,全球产生了约 6 300 万 t 塑料垃圾,其中近八成被堆积在垃圾填埋场或自然环境中[23]。塑料因具有高稳定性和难降解等特性,能够长期存在于环境中,且能够在不同环境介质中迁移,如陆地[24]或河流[25]环境中塑料能够迁移至海洋环境,因此,海洋被认为是塑料一个重要的汇[26]。Jambeck 等[27]研究发现,2010 年 192 个沿海国家产生了 2.75 亿 t 塑料垃圾,其中 480 万~1 270 万 t 进入海洋。环境中塑料受物理、化学和生物等因素影响破碎降解形成小的塑料碎片,2004 年 Rochman 等[28-29]首次将小于 5 mm 的塑料碎片定义为微塑料(MPs)。MPs 按照其来源可以分为初级 MPs 和次级 MPs。初级 MPs 通常是指被生产出来的,进入环境的小尺寸颗粒[30]。Wang 等[30]建立的中国大陆初级 MPs 排放清单显示,2015 年初级 MPs 在中国大陆的排放量高达 737.29 万 t(轮胎粉尘和合成纤维的比例最高,分别占排放总量的 53.91% 和 28.77%),其中六分之一进入了水环境[30]。次级 MPs 是指塑料碎片在环境中受到物理、化学和生物等影响而破碎降解形成的 MPs 碎片[31]。目前在湖泊[32]、河流[33]、海洋[34]、深海[35]、南极[36]和北极[37]等水环境中均检测到 MPs 的存在。此外,在个人护理和化妆品[38]、食盐[39]、啤酒[40]和饮用水[41]等商品中也能检测到 MPs 的存在。MPs 因粒径小等特性易被生物摄入,目前已经在脊椎动物如鱼类[42]、鸟类[43]和哺乳动物[44],以及无脊椎动物如珊瑚[45]和贻贝[46]等生物中检测到 MPs 的存在。MPs 在分子、细胞器、细胞、组织、器官和个体等层面均能对生物体造成不利影响[47](图 4-1),因此,关注 MPs 对生物体的影响具有重要的现实意义。

研究发现,MPs 可以吸附污染物,影响污染物对生物的毒性效应,疏水力和静电力是 MPs 吸附污染物的主要机制,同时环境条件也会影响吸附过程[48]。Guo 等[49]研究表明,MPs 可以作为重金属和多种有机污染物的载体,进而改变污染物对生物存活、生长和繁殖等方面的影响。Cheng 等[50]研究发现,MPs 能够吸附 PFAS 并作为载体成为 PFAS 从水环境向生物群运输的潜在途径。因此,有必要研究 MPs 的存在是否会造成有机污染物对水生生物的影响差异,并探究其对水生生物的毒性作用机制,进而为研究有机污染物和 MPs 的水生态毒性和健康风险提供理论依据。

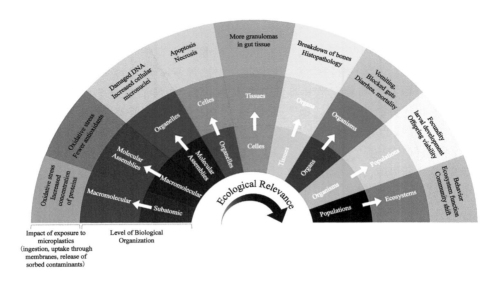

图 4-1　MPs 在生物组织各个层面的潜在毒性影响及其生态相关性[47]

4.2　国内外研究现状

4.2.1　PFOA 毒性研究进展

大量研究显示,PFOA 会对水生生物的存活、生长和繁殖等诸多方面造成负面影响。Weiss-Errico 等[51]发现在 7 d 暴露期内,PFOA 对斑马鱼胚胎的 LC_{50} 维持在 50 mg/L 左右,Satbhai 等[52]发现 PFOA 对斑马鱼胚胎的受精后 24 h 的 LC_{50} 为 82 μmol/L,而 Godfrey 等[53]发现 PFOA 对斑马鱼胚胎 96 h 的 LC_{50} 为 473 mg/L,Wasel 等[54]发现 PFOA 对斑马鱼胚胎的受精后 96 h 的 LC_{50} 为 561.05 mg/L。此外,Yu 等[55]发现 PFOA 会影响早期胚胎阶段斑马鱼的心脏细胞、晶状体细胞、孵化腺细胞及巨噬细胞,进而影响其心跳和运动行为。Dong 等[56]发现 PFOA 会引起雄性鲫鱼(Carassius auratus)的神经毒性并破坏脂质代谢,还会引起过氧化物酶体增殖,从而导致线粒体功能障碍。Ulhaq 等[57]发现 PFOA 会影响斑马鱼幼鱼的行为能力,这可能与 PFOA 的碳链长度和附着官能团有关,同时 PFOA 浓度会影响幼鱼的行为反应。Ulhaq 等[58]还发现斑马鱼在 10 μg/L PFOA 的 40 d 暴露期内,体内的 PFOA 在 20～30 d 内达到稳定浓度,而肠道、胆汁和卵母细胞中积聚的高浓度 PFOA 表明,PFOA 能进行肝、肠循环和母体转移。Tang 等[59]发现 PFOA 会损伤黑斑蛙(*Rana nigromaculata*)肝脏的结构,显著增加 *CYP*3*A*、*Nrf*2 和 *NQO*1 mRNA 的表达水平,在肝脏中引起氧化应激进而造成肝损伤。Shi 等[60]发现 PFOA 可能会损害小鼠肠道屏障的完整性并损伤突触结构,还能通过增加脂多糖、肿瘤坏死因子-α、白细胞介素-1β 和环氧合酶-2 以及减少白细胞介蛋白-10 而引起肠道和大脑炎症。Shi 等[61]还发现 PFOA 会造成肥胖小鼠体重下降、肝脏肿大、行为异常、突触结构受损、神经炎症、神经胶质细胞活化、神经生长因子下降、肠道微生物群改变和血清代谢产物紊乱,这些结果表明 PFOA 能影响肥胖小鼠认知功能。Zhang 等[62]发现 PFOA 能在斑马鱼肾脏中累积,当 PFOA 浓度为 0.05 mg/L 时,79 d 比 21 d 暴露期使得 PFOA 在鱼肾脏中的累积量显著增加 89.14%,同时还发现 PFOA 会通过增加促炎细胞因子来影响抗体,而抗体水平的变化进一步影响了其他细胞因子的表达,最终导致斑马鱼免疫系统紊乱。Gebreab 等[63]发现 PFOA 会影响斑马鱼胚胎发育及胚胎代谢情况,改变三羧酸循环(TCA)相关的几种代谢物以及影响相关的电子传递链和氧化磷酸化,并造成线粒体损伤。Gaballah 等[64]还发现 PFOA 会影响斑马鱼幼鱼的生长发育,并产生神经毒性。Neisiani 等[65]发现 PFOA 会对小鼠大脑特定区域的神经发生和髓鞘形成造成负面影响。Godfrey 等[66]发现 PFOA 能充当甲状腺干扰物,并通过表面活性剂系统改变斑马鱼鱼鳔的正常发育。Jantzen 等[67]发现 PFOA 在亚致死暴露下会对斑马鱼产生永久性形态学、基因表达和行为方面的影响,且这些影响存在持续性。Jantzen 等[68]还发现 PFOA 的长期暴露会导致斑马鱼后代的早期发育被推迟,同时改变斑马鱼有机阴离子转运蛋白基因的表达。Adedara 等[69]发现 PFOA 会影响成年斑马鱼攻击性和焦虑样行为,抑制乙酰胆碱酶(AchE)活性,同时增加三磷酸腺苷(ATP)和一磷酸腺苷(AMP)的水解。Li 等[70]发现 PFOA 能够积聚在贻贝(*Mytiluse dulis*)消化腺中引起氧化应激并诱导贻贝的解毒过程,且 c-JunN-末端激酶(JNK)和 p38 依赖性丝裂原活化蛋白激酶(MAPK)通路、过氧化物酶体增殖物活化受体 γ(PPARγ)通路等解毒相关通路受到 PFOA 影响,同时,PFOA 还会干扰贻贝的脂质代谢、氨基酸代谢和碳水化合物代谢等代谢过程。Wei 等[71]发现 PFOA 会造成生物内分泌紊乱,鳑鱼(*Gobiocypris rarus*)基因组中有 43 个参与多种生物过程的基因受到 PFOA 影响,这些基因参与脂质代谢和运输、激素作用、免疫反应和线粒体功能的生物过程,PFOA 抑制了参与脂肪酸合成和运输及负责甲状腺激素生物合成的基因,同时诱导了雌激素反应及与胆固醇细胞运输相关的基因。因此,研究 PFOA 的生物健康风险具有重要的现实意义。

4.2.2 MPs 毒性研究进展

目前,全球范围内水环境中已检测到大量聚乙烯(PE)、聚丙烯(PP)、聚氯乙烯(PVC)、聚苯乙烯(PS)、聚酰胺(PA)和聚对苯二甲酸乙二醇酯(PET)等 MPs 的存在,MPs 对水生生物负面影响的研究已较为充分,不同尺寸 MPs 对生物体的影响存在差异。Lu 等[72]将 5 月龄斑马鱼暴露在 20 mg/L PS 7 d 后,发现 5 μm PS 在鱼鳃、肠道和肝脏中累积,20 μm PS 在鳃和肠道中累积。将斑马鱼暴露于 70 nm 和 5 μm PS 21 d 后,PS 能够诱导斑马鱼氧化应激,造成肝脏代谢谱改变,并干扰能量和脂质代谢。van Pomeren 等[73]发现 25 nm 和 50 nm 的 PS 能被斑马鱼摄取,最终可积聚在其眼睛中。比较不同尺寸 PE 碎片[(17.23±3.43) μm 和 (34.43±13.09) μm] 和微珠(40~48 μm)对大型蚤(*Daphnia magna*)影响的一项研究发现,较小 PE 碎片的 6 d 生物累积程度显著高于较大 PE,且 21 d 暴露期后较小 PE 对大型蚤的生长、存活和繁殖造成的负面影响显著高于较大尺寸 PE[74]。Jeong 等[75]将轮虫(*Brachionus koreanus*)暴露于 0.05 μm、0.5 μm 和 6 μm PS 中,发现较小的 PS 对轮虫毒性更大。Gu 等[76]将斑马鱼暴露于 100 nm、5 μm 和 200 μm PS 21 d 后,发现 MPs 对生物的影响具有尺寸依赖性,会造成斑马鱼肠道中的免疫细胞功能障碍,引起肠道内致病菌丰度增加,同时,基因表达分析揭示了三种尺寸 PS 作用的相似性和差异性。Zhang 等[77]发现不同粒径 PS 对海洋青鳉(*Oryzias melastigma*)代谢的影响存在差异。Kang 等[78]发现纳米级 PS(50 nm)对海洋青鳉具有更明显的氧化应激,而微米级 PS(45 μm)对海洋青鳉肠道微生物群组成的影响更大。Sun 等[79]研究发现,0.07 μm PS 比 0.7 μm 和 7 μm PS 对轮虫(*Brachionus plicatilis*)的生长和繁殖产生更明显的负面影响。

4.2.3 污染物与 MPs 复合毒性研究进展

环境中污染物较少以单一形式存在,通常是多种污染物共存产生协同作用、相加作用或拮抗作用等复合影响。MPs 可以通过吸附污染物等方式影响污染物对生物的毒性效应。研究发现,MPs 与重金属、抗生素、药物和有机污染物等共存时对生物体的毒性作用存在差异。银纳米塑料和 20 nm PS 共存时对淡水藻类(*Chlamydomonas reinhardtii* 和 *Ochromonas danica*)造成的损伤高于单一污染物[80]。PS 与重金属铅、铜、镉和镍的联合作用对大型蚤存活的影响随着 PS 浓度的增加由拮抗转变为相加[81]。对于 MPs 和有机污染物的联合毒性,Zhou 等[82]发现 500 nm PS 能显著增加抗生素(土霉素和氟苯尼考)对成年血蛤(*Tegillarca granosa*)在氧化应激、免疫反应和基因水平方面的毒性影响。

Zhang 等[83]发现 0.1 μm PS 可增加罗红霉素在红罗非鱼(Oreochromis niloticus)中的生物累积,还能减轻罗红霉素诱导的神经毒性和氧化损伤。Shi 等[84]发现 500 nm PS 与舍曲林间存在协同毒性,当它们共同暴露时会显著影响血蛤的免疫毒性。Brandts 等[85]发现(110±6.9) nm PS 降低了卡马西平对贻贝(Mytilus galloprovincialis)的毒性,并激活了不同的反应机制。Tang 等[86]发现双酚 A 与(490±11) nm PS 共存对血蛤的免疫毒性和神经毒性显著高于单一污染物。Yang 等[87]发现 5 μm PS 和 6∶2 氯化多氟醚磺酸盐对斑马鱼幼鱼的氧化应激和炎症反应具有明显的协同毒性。同时,Trevisan 等[88]发现纳米级 PS 可以降低多环芳烃引起的斑马鱼发育畸形和血管发育受损。O'Donovan 等[89]发现 PE(11~13 μm)与苯并[a]芘或 PE 与 PFOS 共同暴露后,会对蛤蜊(Scrobicularia plana)消化腺和鳃组织造成氧化损伤。Le Bihanic 等[90]发现 PE 和 PFOS 共存会降低海洋青鳉的胚胎存活率和孵化率,而 PE 与苯并[a]芘或与二苯甲酮-3 共存对海洋青鳉的影响高于单一污染物,包括降低幼鱼生长、增加异常发育和异常行为。Hanachi 等[91]发现 PS 碎片对虹鳟鱼(Onchorhynchus mykiss)氨基酸和脂肪酸组成的影响最小,对鱼肌肉蛋白质含量没有影响;然而,PS 和毒死蜱共存会造成氨基酸和脂肪酸组成以及蛋白质含量的显著变化。Li 等[92]发现 PS 与邻苯二甲酸二丁酯共存对桡足类(Tigriopus japonicus)的繁殖具有拮抗作用。因此,MPs 与其他污染物共存对生物造成的直接影响或潜在影响值得进一步关注。

4.3 研究内容及意义

PFOA 对水生生物造成的健康风险已经得到证实[93],而 PS 作为一种广泛存在于环境中的污染物已在淡水野生鱼类中被检测到[94-95]。研究发现,暴露于环境中的 MPs 可以吸附 PFOA[96],PFOA 与 MPs 可通过食物链在生物体中传递[50]。斑马鱼作为一种应用广泛的水生生物研究模型[97],具有体型小、繁殖周期短、繁殖成本低、易观察、基因组背景清晰等特点[98]。Collins 等[99]提供了 26 000 多个基因的全面注释;Howe 等[100]还发现斑马鱼基因组与人类基因组具有高度的相似性,约七成人类基因至少有一个明显的斑马鱼直系同源物,因此,斑马鱼基因层面的研究为污染物对人类的影响提供了参考。尽管已有研究开展了 PFOA 或 PS 对斑马鱼的影响研究,但两种污染物共同存在对斑马鱼的影响研究还不够充分。因此,开展 PFOA 和 PS 对斑马鱼影响的研究具有重要意义。

本研究选取环境中广泛存在的有机污染物 PFOA 和 PS-MPs 作为目标污染物,以典型模式生物斑马鱼为受试生物,通过急性毒性实验、PFOA 含量测定、生物富集分析、组织病理学分析以及转录组测序等实验手段研究 PS 对 PFOA 斑马鱼毒性效应的影响以及不同粒径 PS 对 PFOA 的斑马鱼毒性差异。

研究整体思路如下：对 PFOA 和 PS 分别设置适当浓度梯度进行 96 h 急性毒性实验，计算半数致死浓度（LC_{50}）；随后固定 PS 浓度，以 PFOA 浓度为变量进行急性毒性实验，并计算 LC_{50}；通过液相色谱/质谱联用技术测定 PS 存在/不存在时 PFOA 在斑马鱼和溶液中的含量并计算 PFOA 的生物富集系数；通过组织病理学分析研究 PS 存在/不存在时 PFOA 对斑马鱼鳃、肠和肌肉组织造成的影响；随后通过转录组测序分析 PS 存在/不存在时 PFOA 对斑马鱼基因表达的影响，通过筛选差异表达基因，并对差异表达基因进行 GO 富集分析和通路富集分析，研究 PS 存在/不存在时 PFOA 对斑马鱼的毒性机制。以转录组测序为基础，以 PFOA 暴露组为对照，分析 MPs 与 PFOA 共同暴露时不同粒径 MPs 对 PFOA 生物毒性的影响，筛选差异表达基因，并对差异表达基因进行 GO 富集分析和通路富集分析，研究不同尺寸 MPs 对斑马鱼影响的毒性机制。

具体研究技术路线如图 4-2 所示。本研究评估了 PS 存在/不存在情况下 PFOA 对斑马鱼的影响，比较了 PFOA 与不同粒径 MPs 共存时对斑马鱼的影响，为水环境中 PFAS 与 MPs 对水生生物的复合影响及 MPs 粒径对 PFAS 生物毒性影响的研究提供理论参考。

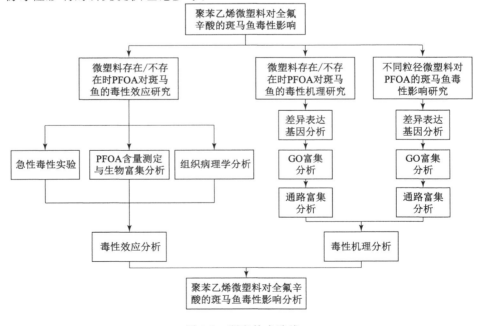

图 4-2　研究技术路线

参考文献

[1] AHRENS L, BUNDSCHUH M. Fate and effects of poly-and perfluoroalkyl substances in the aquatic environment: a review[J]. Environmental toxicology and chemistry, 2014, 33(9): 1921-1929.

[2] RANKIN K, MABURY S A, JENKINS T M, et al. A North American and global survey of perfluoroalkyl substances in surface soils: distribution patterns and mode of occurrence[J]. Chemosphere, 2016, 161: 333-341.

[3] ZAREITALABAD P, SIEMENS J, HAMER M, et al. Perfluorooctanoic acid (PFOA) and perfluorooctanesulfonic acid (PFOS) in surface waters, sediments, soils and wastewater: a review on concentrations and distribution coefficients[J]. Chemosphere, 2013, 91(6): 725-732.

[4] AHRENS L, HARNER T, SHOEIB M, et al. Improved characterization of gas-particle partitioning for per-and polyfluoroalkyl substances in the atmosphere using annular diffusion denuder samplers[J]. Environmental science and technology, 2012, 46(13): 7199-7206.

[5] MUDUMBI J B N, NTWAMPE S K O, MATSHA T, et al. Recent developments in polyfluoroalkyl compounds research: a focus on human/environmental health impact, suggested substitutes and removal strategies[J]. Environmental monitoring and assessment, 2017, 189(8): 402.

[6] LINDSTROM A B, STRYNAR M J, LIBELO E L. Polyfluorinated compounds: past, present, and future [J]. Environmental science and technology, 2011, 45(19): 7954-7961.

[7] POPR. Proposal to list pentadecafluorooctanoic acid, its salts and PFOA-related compounds in annexes A, B and/or C to the stockholm convention on persistent organic pollutants[R]. Stockholm Conventions, 2015.

[8] POPR. Pentadecafluorooctanoic acid, its salts and PFOA-related compounds [R]. Stockholm Conventions, 2016.

[9] GEBBINK W A, VAN LEEUWEN S P J. Environmental contamination and human exposure to PFASs near a fluorochemical production plant: review of historic and current PFOA and GenX contamination in the Netherlands [J]. Environment international, 2020, 137: 105583.

[10] QUIÑONES O, SNYDER S A. Occurrence of perfluoroalkyl carboxylates

and sulfonates in drinking water utilities and related waters from the United States[J].Environmental science and technology,2009,43(24):9089-9095.

[11] SUNGUR Ş.Dietary exposure to perfluorooctanoic acid (PFOA) and perfluorooctane sulfonic acid (PFOS):a review of recent literature[J].Toxin reviews,2018,37(2):106-116.

[12] ANDREWS D Q,NAIDENKO O V.Population-wide exposure to per-and polyfluoroalkyl substances from drinking water in the United States[J].Environmental science and technology letters,2020,7(12):931-936.

[13] MORIKAWA A,KAMEI N Y,HARADA K,et al.The bioconcentration factor of perfluorooctane sulfonate is significantly larger than that of perfluorooctanoate in wild turtles (*Trachemys scripta elegans* and *Chinemys Reevesii*):an AI river ecological study in Japan[J].Ecotoxicology and environmental safety,2006,65(1):14-21.

[14] OLIVERO-VERBEL J,TAO L,JOHNSON-RESTREPO B,et al.Perfluorooctanesulfonate and related fluorochemicals in biological samples from the north coast of Colombia[J].Environmental pollution,2006,142(2):367-372.

[15] CALAFAT A M,NEEDHAM L L,KUKLENYIK Z,et al.Perfluorinated chemicals in selected residents of the American continent[J].Chemosphere,2006,63(3):490-496.

[16] GYLLENHAMMAR I,BENSKIN J P,SANDBLOM O,et al.Perfluoroalkyl acids (PFAAs) in children's serum and contribution from PFAA-contaminated drinking water[J].Environmental science and technology,2019,53(19):11447-11457.

[17] EPA.Drinking water health advisory for perfluorooctanoic acid(PFOA)[R].National Service Center for Environmental Publications(NSCEP),2016.

[18] EPA.Drinking water health advisory for perfluorooctane sulfonate(PFOS)[R].National Service Center for Environmental Publications(NSCEP),2016.

[19] POST G B.Recent US state and federal drinking water guidelines for per- and polyfluoroalkyl substances[J].Environmental toxicology and chemistry,2021,40(3):550-563.

[20] LOOS R,LOCORO G,HUBER T,et al.Analysis of perfluorooctanoate (PFOA) and other perfluorinated compounds (PFCs) in the River Po watershed in N-Italy[J].Chemosphere,2008,71(2):306-313.

[21] SO M K, MIYAKE Y, YEUNG W Y, et al. Perfluorinated compounds in the Pearl River and Yangtze River of China[J]. Chemosphere, 2007, 68(11):2085-2095.

[22] WEI S, CHEN L Q, TANIYASU S, et al. Distribution of perfluorinated compounds in surface seawaters between Asia and Antarctica[J]. Marine pollution bulletin, 2007, 54(11):1813-1818.

[23] GEYER R, JAMBECK J R, LAW K L. Production, use, and fate of all plastics ever made[J]. Science advances, 2017, 3(7):1700782.

[24] JAMBECK J R, GEYER R, WILCOX C, et al. Plastic waste inputs from land into the ocean[J]. Science, 2015, 347(6223):768-771.

[25] LEBRETON L C M, VAN DER ZWET J, DAMSTEEG J W, et al. River plastic emissions to the world's oceans[J]. Nature communications, 2017, 8:15611.

[26] SUARIA G, AVIO C G, MINEO A, et al. The Mediterranean Plastic Soup: synthetic polymers in Mediterranean surface waters[J]. Scientific reports, 2016, 6:37551.

[27] JAMBECK J R, GEYER R, WILCOX C, et al. Plastic waste inputs from land into the ocean[J]. Science, 2015, 347(6223):768-771.

[28] ROCHMAN C M. Microplastics research-from sink to source[J]. Science, 2018, 360(6384):28-29.

[29] THOMPSON R C, OLSEN Y, MITCHELL R P, et al. Lost at sea: where is all the plastic? [J]. Science, 2004, 304(5672):838.

[30] WANG T, LI B J, ZOU X Q, et al. Emission of primary microplastics in mainland China: invisible but not negligible[J]. Water Research, 2019, 162:214-224.

[31] TONG H Y, ZHONG X C, DUAN Z H, et al. Micro-and nanoplastics released from biodegradable and conventional plastics during degradation: formation, aging factors, and toxicity[J]. Science of the total environment, 2022, 833:155275.

[32] ERIKSEN M, MASON S, WILSON S, et al. Microplastic pollution in the surface waters of the Laurentian Great Lakes[J]. Marine pollution bulletin, 2013, 77(1/2):177-182.

[33] MEIJER L J J, VAN EMMERIK T, VAN DER ENT R, et al. More than 1 000 rivers account for 80% of global riverine plastic emissions into the

ocean[J].Science advances,2021,7(18):5803.

[34] COLE M,LINDEQUE P,HALSBAND C,et al.Microplastics as contaminants in the marine environment:a review[J].Marine pollution bulletin, 2011,62(12):2588-2597.

[35] WOODALL L C,SANCHEZ-VIDAL A,CANALS M,et al.The deep sea is a major sink for microplastic debris[J].Royal society open science, 2014,1(4):140317.

[36] LUSHER A L,TIRELLI V,O'CONNOR I,et al.Microplastics in Arctic polar waters:the first reported values of particles in surface and sub-surface samples[J].Scientific reports,2015,5:14947.

[37] WALLER C L,GRIFFITHS H J,WALUDA C M,et al.Microplastics in the Antarctic marine system:an emerging area of research[J].Science of the total environment,2017,598:220-227.

[38] GUERRANTI C,MARTELLINI T,PERRA G,et al.Microplastics in cosmetics:environmental issues and needs for global bans[J].Environmental toxicology and pharmacology,2019,68:75-79.

[39] YANG D Q,SHI H H,LI L,et al.Microplastic pollution in table salts from China[J].Environmental science and technology,2015,49(22):13622-13627.

[40] BERTHILLER F.Foreword[J].Food additives and contaminants:Part A, 2010,27(5):575.

[41] PIVOKONSKY M,CERMAKOVA L,NOVOTNA K,et al.Occurrence of microplastics in raw and treated drinking water[J].Science of the total environment,2018,643:1644-1651.

[42] FOEKEMA E M,DE GRUIJTER C,MERGIA M T,et al.Plastic in north sea fish[J].Environmental science and technology,2013,47(15):8818-8824.

[43] CODINA-GARCÍA M,MILITÃO T,MORENO J,et al.Plastic debris in Mediterranean seabirds[J]. Marine pollution bulletin, 2013, 77(1/2): 220-226.

[44] LUSHER A L,HERNANDEZ-MILIAN G,O'BRIEN J,et al.Microplastic and macroplastic ingestion by a deep diving, oceanic cetacean: the True's beaked whale *Mesoplodon mirus*[J].Environmental pollution,2015,199:185-191.

[45] HALL N M,BERRY K L E,RINTOUL L,et al.Microplastic ingestion by scleractinian corals[J].Marine biology,2015,162(3):725-732.

[46] LI J N,QU X Y,SU L,et al.Microplastics in mussels along the coastal

waters of China[J]. Environmental pollution, 2016, 214:177-184.

[47] KHALID N, AQEEL M, NOMAN A, et al. Linking effects of microplastics to ecological impacts in marine environments[J]. Chemosphere, 2021, 264(2):128541.

[48] TOURINHO P S, KOCÍ V, LOUREIRO S, et al. Partitioning of chemical contaminants to microplastics: Sorption mechanisms, environmental distribution and effects on toxicity and bioaccumulation[J]. Environmental pollution, 2019, 252:1246-1256.

[49] GUO X, WANG J L. The chemical behaviors of microplastics in marine environment: a review[J]. Marine pollution bulletin, 2019, 142:1-14.

[50] CHENG Y, MAI L, LU X W, et al. Occurrence and abundance of poly-and perfluoroalkyl substances (PFASs) on microplastics (MPs) in Pearl River Estuary (PRE) region: spatial and temporal variations[J]. Environmental pollution, 2021, 281:117025.

[51] WEISS-ERRICO M J, BERRY J P, O'SHEA K E. β-cyclodextrin attenuates perfluorooctanoic acid toxicity in the zebrafish embryo model[J]. Toxics, 2017, 5(4):31.

[52] SATBHAI K, VOGS C, CRAGO J. Comparative toxicokinetics and toxicity of PFOA and its replacement GenX in the early stages of zebrafish[J]. Chemosphere, 2022, 308(1):136131.

[53] GODFREY A, ABDEL-MONEIM A, SEPÚLVEDA M S. Acute mixture toxicity of halogenated chemicals and their next generation counterparts on zebrafish embryos[J]. Chemosphere, 2017, 181:710-712.

[54] WASEL O, THOMPSON K M, FREEMAN J L. Assessment of unique behavioral, morphological, and molecular alterations in the comparative developmental toxicity profiles of PFOA, PFHxA, and PFBA using the zebrafish model system[J]. Environment international, 2022, 170:107642.

[55] YU J, CHENG W Q, JIA M, et al. Toxicity of perfluorooctanoic acid on zebrafish early embryonic development determined by single-cell RNA sequencing[J]. Journal of hazardous materials, 2022, 427:127888.

[56] DONG H K, LU G H, YAN Z H, et al. Molecular and phenotypic responses of male crucian carp (*Carassius auratus*) exposed to perfluorooctanoic acid[J]. Science of the total environment, 2019, 653:1395-1406.

[57] ULHAQ M, ÖRN S, CARLSSON G, et al. Locomotor behavior in zebrafish (*Danio rerio*) larvae exposed to perfluoroalkyl acids[J]. Aquatic toxicology,

2013,144/145:332-340.

[58] ULHAQ M,SUNDSTRÖM M,LARSSON P,et al.Tissue uptake,distribution and elimination of (14)C-PFOA in zebrafish (*Danio rerio*)[J]. Aquatic toxicology,2015,163:148-157.

[59] TANG J,JIA X Y,GAO N N,et al.Role of the Nrf2-ARE pathway in perfluorooctanoic acid (PFOA)-induced hepatotoxicity in *Rana nigromaculata*[J].Environmental pollution,2018,238:1035-1043.

[60] SHI L C,ZHENG J J,YAN S K,et al.Exposure to perfluorooctanoic acid induces cognitive deficits via altering gut microbiota composition,impairing intestinal barrier integrity,and causing inflammation in gut and brain[J].Journal of agricultural and food chemistry,2020,68(47):13916-13928.

[61] SHI L C,DENG X,LIU X N,et al.The effect of chronic exposure to a low concentration of perfluorooctanoic acid on cognitive function and intestinal health of obese mice induced by a high-fat diet[J].Food and chemical toxicology:an international journal published for the british industrial biological research association,2022,168:113395.

[62] ZHANG H J,SHEN L L,FANG W D,et al.Perfluorooctanoic acid-induced immunotoxicity via NF-kappa B pathway in zebrafish (*Danio rerio*) kidney[J]. Fish and shellfish immunology,2021,113:9-19.

[63] GEBREAB K Y,EEZA M N H,BAI T Y,et al.Comparative toxicometabolomics of perfluorooctanoic acid (PFOA) and next-generation perfluoroalkyl substances[J].Environmental pollution,2020,265(1):114928.

[64] GABALLAH S,SWANK A,SOBUS J R,et al.Evaluation of developmental toxicity,developmental neurotoxicity,and tissue dose in zebrafish exposed to GenX and other PFAS[J].Environmental health perspectives,2020,128(4):47005.

[65] NEISIANI A K,MOUSAVI M K,SOLTANI M,et al.Perfluorooctanoic acid exposure and its neurodegenerative consequences in C57BL6/J mice [J].Naunyn-Schmiedeberg's archives of pharmacology,2023,396(10):2357-2367.

[66] GODFREY A,HOOSER B,ABDELMONEIM A,et al.Thyroid disrupting effects of halogenated and next generation chemicals on the swim bladder development of zebrafish[J].Aquatic toxicology,2017,193:228-235.

[67] JANTZEN C E,ANNUNZIATO K M,COOPER K R.Behavioral,morphomet-

ric,and gene expression effects in adult zebrafish (*Danio rerio*) embryonically exposed to PFOA, PFOS, and PFNA[J]. Aquatic toxicology, 2016, 180: 123-130.

[68] JANTZEN C E,TOOR F,ANNUNZIATO K A,et al.Effects of chronic perfluorooctanoic acid (PFOA) at low concentration on morphometrics,gene expression,and fecundity in zebrafish (*Danio rerio*)[J].Reproductive toxicology, 2017,69:34-42.

[69] ADEDARA I A,SOUZA T P,CANZIAN J,et al.Induction of aggression and anxiety-like responses by perfluorooctanoic acid is accompanied by modulation of cholinergic-and purinergic signaling-related parameters in adult zebrafish[J]. Ecotoxicology and environmental safety,2022,239:113635.

[70] LI F L,YU Y X,GUO M M,et al.Integrated analysis of physiological, transcriptomics and metabolomics provides insights into detoxication disruption of PFOA exposure in *Mytilus edulis*[J].Ecotoxicology and environmental safety,2021,214:112081.

[71] WEI Y H,LIU Y,WANG J S,et al.Toxicogenomic analysis of the hepatic effects of perfluorooctanoic acid on rare minnows (*Gobiocypris rarus*) [J].Toxicology and applied pharmacology,2008,226(3):285-297.

[72] LU Y F,ZHANG Y,DENG Y F,et al.Uptake and accumulation of polystyrene microplastics in zebrafish (*Danio rerio*) and toxic effects in liver [J].Environmental science and technology,2016,50(7):4054-4060.

[73] VAN POMEREN M,BRUN N R,PEIJNENBURG W J G M,et al.Exploring uptake and biodistribution of polystyrene (nano)particles in zebrafish embryos at different developmental stages[J].Aquatic toxicology,2017,190:40-45.

[74] AN D,NA J,SONG J,et al.Size-dependent chronic toxicity of fragmented polyethylene microplastics to *Daphnia Magna*[J].Chemosphere,2021, 271:129591.

[75] JEONG C B,WON E J,KANG H M,et al.Microplastic size-dependent toxicity,oxidative stress induction,and p-JNK and p-p38 activation in the monogonont rotifer (brachionus koreanus)[J].Environmental science and technology,2016,50(16):8849-8857.

[76] GU W Q,LIU S,CHEN L,et al.Single-cell RNA sequencing reveals size-dependent effects of polystyrene microplastics on immune and secretory cell populations from zebrafish intestines[J]. Environmental science and

technology,2020,54(6):3417-3427.

[77] ZHANG X,WEN K,DING D X,et al.Size-dependent adverse effects of microplastics on intestinal microbiota and metabolic homeostasis in the marine medaka (*Oryzias melastigma*)[J].Environment international, 2021,151:106452.

[78] KANG H M,BYEON E,JEONG H,et al.Different effects of nano-and microplastics on oxidative status and gut microbiota in the marine medaka *Oryzias melastigma*[J].Journal of hazardous materials,2021,405:124207.

[79] SUN Y F,XU W J,GU Q J,et al.Small-sized microplastics negatively affect rotifers:changes in the key life-history traits and rotifer-Phaeocystis population dynamics[J].Environmental science and technology,2019,53(15):9241-9251.

[80] HUANG B,WEI Z B,YANG L Y,et al.Combined toxicity of silver nanoparticles with hematite or plastic nanoparticles toward two freshwater algae[J].Environmental science and technology,2019,53(7):3871-3879.

[81] YUAN W K,ZHOU Y F,CHEN Y L,et al.Toxicological effects of microplastics and heavy metals on the *Daphnia Magna*[J].Science of the total environment,2020,746:141254.

[82] ZHOU W S,TANG Y,DU X Y,et al.Fine polystyrene microplastics render immune responses more vulnerable to two veterinary antibiotics in a bivalve species[J].Marine pollution bulletin,2021,164:111995.

[83] ZHANG S S,DING J N,RAZANAJATOVO R M,et al.Interactive effects of polystyrene microplastics and roxithromycin on bioaccumulation and biochemical status in the freshwater fish red tilapia (*Oreochromis niloticus*)[J]. Science of the total environment,2019,648:1431-1439.

[84] SHI W,HAN Y,SUN S G,et al.Immunotoxicities of microplastics and sertraline,alone and in combination,to a bivalve species:size-dependent interaction and potential toxication mechanism[J].Journal of hazardous materials,2020,396:122603.

[85] BRANDTS I,TELES M,GONÇALVES A P,et al.Effects of nanoplastics on Mytilus galloprovincialis after individual and combined exposure with carbamazepine[J].Science of the total environment,2018,643:775-784.

[86] TANG Y,ZHOU W S,SUN S G,et al.Immunotoxicity and neurotoxicity of bisphenol A and microplastics alone or in combination to a bivalve species,*Tegil-*

larca granosa[J].Environmental pollution,2020,265(1):115115.

[87] YANG H L,LAI H,HUANG J,et al.Polystyrene microplastics decrease F-53B bioaccumulation but induce inflammatory stress in larval zebrafish[J].Chemosphere,2020,255:127040.

[88] TREVISAN R,VOY C,CHEN S X,et al.Nanoplastics decrease the toxicity of a complex PAH mixture but impair mitochondrial energy production in developing zebrafish[J].Environmental science and technology,2019,53(14):8405-8415.

[89] O'DONOVAN S,MESTRE N C,ABEL S,et al.Ecotoxicological effects of chemical contaminants adsorbed to microplastics in the clam *Scrobicularia Plana*[J].Frontiers in marine science,2018,5:143.

[90] LE BIHANIC F,CLÉRANDEAU C,CORMIER B,et al.Organic contaminants sorbed to microplastics affect marine medaka fish early life stages development[J].Marine pollution bulletin,2020,154:111059.

[91] HANACHI P,KARBALAEI S,YU S J.Combined polystyrene microplastics and chlorpyrifos decrease levels of nutritional parameters in muscle of rainbow trout (*Oncorhynchus mykiss*)[J].Environmental science and pollution research,2021,28(45):64908-64920.

[92] LI Z C,ZHOU H,LIU Y,et al.Acute and chronic combined effect of polystyrene microplastics and dibutyl phthalate on the marine copepod *Tigriopus japonicus*[J].Chemosphere,2020,261:127711.

[93] PULSTER E L,WICHTERMAN A E,SNYDER S M,et al.Detection of long chain per-and polyfluoroalkyl substances (PFAS) in the benthic Golden tilefish (*Lopholatilus chamaeleonticeps*) and their association with microscopic hepatic changes[J].Science of the total environment,2022,809:151143.

[94] ANDRADY A L,NEAL M A.Applications and societal benefits of plastics[J].Philosophical transactions of the royal society of London Series B,biological sciences,2009,364(1526):1977-1984.

[95] CERA A,SIGHICELLI M,SODO A,et al.Microplastics distribution and possible ingestion by fish in lacustrine waters (Lake Bracciano,Italy)[J].Environmental science and pollution research,2022,29(45):68179-68190.

[96] SCOTT J W,GUNDERSON K G,GREEN L A,et al.Perfluoroalkylated substances (PFAS) associated with microplastics in a lake environment

[J].Toxics,2021,9(5):106.

[97] CHAKRABORTY C,SHARMA A R,SHARMA G,et al.Zebrafish: a complete animal model to enumerate the nanoparticle toxicity[J].Journal of nanobiotechnology,2016,14(1):65.

[98] DAI Y J,JIA Y F,CHEN N,et al.Zebrafish as a model system to study toxicology[J].Environmental toxicology and chemistry,2014,33(1):11-17.

[99] COLLINS J E,WHITE S,SEARLE S M J,et al.Incorporating RNA-seq data into the zebrafish Ensembl genebuild[J].Genome research,2012,22(10):2067-2078.

[100] HOWE K,CLARK M D,TORROJA C F,et al.The zebrafish reference genome sequence and its relationship to the human genome[J].Nature,2013,496:49.

第 5 章　聚苯乙烯微塑料对全氟辛酸斑马鱼毒性效应影响研究

5.1　引言

研究发现,水环境可作为 PFOA 与 MPs 重要的汇,随着 PFOA 与 MPs 的环境丰度不断增加,这两种污染物进入生物体及对生物造成不利影响的风险不断增加。本研究选用斑马鱼作为受试对象,以 PFOA 与不同粒径聚苯乙烯微塑料作为实验材料,通过急性毒性实验、生物富集分析和组织病理学分析的方法研究 PS-MPs 存在/不存在情况下 PFOA 对斑马鱼的毒性效应。

5.2　实验材料与方法

5.2.1　受试生物

选择 2~3 月龄斑马鱼作为实验对象。将大约 1 000 条斑马鱼均匀分布在 5 个水族箱中,用曝气自来水驯养一周以上。驯养及后续实验严格按照 OECD(Organization for Economic Cooperation and Development)指南进行[1]。驯养期间控制曝气自来水 pH 值为 7~7.5,溶解氧浓度>80%,电导率≤10 μS/cm,盐度<0.2‰,温度为(24±1) ℃,光暗周期为 14 h 光照∶10 h 黑暗。驯养期间每天喂食两次卤虾(*Artemia nauplii*),并及时取出死亡斑马鱼。驯养过程中,斑马鱼死亡数量不超过总数的 5%。在实验开始前,提前 24 h 停止喂食。随机选取 10 条斑马鱼测得体长与体重分别为(25.3±3.2) mm 和 (0.105±0.022) g。

5.2.2 试剂与仪器

全氟辛酸(PFOA,$C_8HF_{15}O_2$,CAS:335-67-1)购自上海阿拉丁生化科技股份有限公司,PFOA纯度>96%。浓度为2.5%v/v三种的不同尺寸的单分散聚苯乙烯微球(PS-MPs,直径分别为0.2 μm、2 μm和20 μm,记作0.2 PS、2 PS和20 PS)购自倍思乐色谱技术开发中心。4%多聚甲醛固定剂(PFA,CAS:30525-89-4)购自Macklin(中国)。

本章所用实验仪器见表5-1。

表5-1 实验仪器

仪器名称	仪器型号	生产厂家
超声波清洗器	KH-500DE	昆山禾创超声仪器有限公司
液氮罐	YDS-30-125-F	OLABO
超低温冰箱	DW-86L100	青岛海尔生物医疗股份有限公司
超纯水系统	OKP-10	上海涞科仪器有限公司
电子天平	FA2204N	上海菁海仪器有限公司
洁净工作台	DL-CJ-2ND	北京东联哈尔仪器制造有限公司
水质检测仪	HI9829-04	Hanna Instruments
智能人工气候箱	RXZ-1500c-2	宁波江南仪器厂
高效液相色谱-串联质谱仪	ACQUITY UPLC Xevo TQ	Waters
超低温冰箱	DW-86W100	青岛海尔生物医疗股份有限公司
离心机	5424R	SIGMA laborzentrifugen
组织包埋机	HistoCore Arcadia H	Leica
全自动轮转式切片机	HistoCore NANOCUT R	Leica
封闭式组织脱水机	ASP300S	Leica
多功能染色机	ST5020-CV5030	Leica
数字切片扫描仪	MIDI	Pannoramic

5.2.3 急性毒性实验

为了研究PS-MPs对全氟辛酸的斑马鱼急性毒性影响,设置了不同的暴露组研究PFOA、PS(0.2 PS、2 PS和20 PS)及其混合暴露96 h后对斑马鱼的存活影响。对于单一污染物,除空白对照外,设置6个PFOA浓度组和8个PS-MPs(0.2 PS、2 PS和20 PS)浓度组,浓度设置详见表5-2。

第 5 章 聚苯乙烯微塑料对全氟辛酸斑马鱼毒性效应影响研究

表 5-2 单一污染物 96 h 急性毒性实验浓度

PFOA/(mg/L)	0.2 PS/(mg/L)	2 PS/(mg/L)	20 PS/(mg/L)
0	0	0	0
10	40	20	40
20	80	40	80
40	160	80	600
60	240	133	689
80	288	160	792
100	346	192	909
	415	230	1 045
	498	276	1 200

为了评估 PS 对 PFOA 的斑马鱼急性毒性影响,进行了不同浓度的 PFOA 与固定浓度的 PS(40 mg/L 的 0.2 PS、2 PS 和 20 PS)混合毒性实验,PFOA 浓度为 10 mg/L、20 mg/L、40 mg/L、60 mg/L、80 mg/L 和 160 mg/L,浓度设置详见表 5-3。实验开始前,对单一或混合污染物进行超声处理,以确保在不使用任何助溶剂的情况下,各暴露组中污染物完全溶解在水中。随机选取 10 条斑马鱼放入各暴露组的溶液中,并进行 3 次生物学重复。实验过程中不进行喂食或换水,控制温度为(24±1) ℃,光暗周期为 14 h 光照∶10 h 黑暗。每 24 h 对斑马鱼的存活情况进行记录,并及时取出死亡斑马鱼。96 h 暴露后,采用 Probit 法(SPSS 21.0)计算不同暴露组中污染物对斑马鱼的半数致死浓度(LC_{50})。

表 5-3 PS 存在/不存在时 PFOA 的 96 h 急性毒性实验浓度

PFOA/(mg/L)	PFOA+0.2 PS/(mg/L)	PFOA+2 PS/(mg/L)	PFOA+20 PS/(mg/L)
0	0+40	0+40	0+40
10	10+40	10+40	10+40
20	20+40	20+40	20+40
40	40+40	40+40	40+40
60	60+40	60+40	60+40
80	80+40	80+40	80+40
160	160+40	160+40	160+40

5.2.4 PFOA 含量测定与生物富集分析

根据急性毒性实验结果选择 10 mg/L、20 mg/L、40 mg/L、60 mg/L 和 80 mg/L 的 PFOA 与 40 mg/L 的 PS(0.2 PS,2 PS 和 20 PS)混合,来评估 PFOA 在斑马鱼体内的富集情况,浓度设置详见表 5-4。

表 5-4 PS 存在/不存在时 PFOA 含量测定实验浓度

PFOA/(mg/L)	PFOA+0.2 PS/(mg/L)	PFOA+2 PS/(mg/L)	PFOA+20 PS/(mg/L)
0	0+0	0+0	0+0
10	10+40	10+40	10+40
20	20+40	20+40	20+40
40	40+40	40+40	40+40
60	60+40	60+40	60+40
80	80+40	80+40	80+40

与急性毒性实验类似,随机选择 10 条健康斑马鱼放入各暴露组中,并进行 3 次生物学重复,实验过程中不进行喂食或换水,控制温度为 (24 ± 1) ℃,光暗周期为 14 h 光照∶10 h 黑暗。暴露 96 h 后,收集斑马鱼和溶液用于 PFOA 含量测定。对斑马鱼样品进行预处理,称量 0.2 g 样品进行实验。把样品放进 50 mL 离心管并添加 400 μL 内标混合使用溶液(1 μg/L $^{13}C_4$-PFOA 和 5 μg/L 1,2,3,4-$^{13}C_4$-PFOS),加入 5 mL 水,涡旋 1 min;加 10 mL 色谱级乙腈和 30 μL 盐酸,摇动 10 min;加入 2 g 氯化钠,摇动 10 min,以 5 000 r/min 的转速离心 10 min(SIGMA laborzentrifugen);用移液管将上层乙腈溶液移入另一试管中,在 45 ℃水浴中吹氮气至 4 mL 左右;将 100 mL 乙二胺丙酯固相吸附剂、40 mL C_{18} 吸附剂和 20 mL 石墨化炭黑吸附剂放置于 15 mL 试管中摇动 10 min;以 5 000 r/min 的转速离心 10 min,将上层清液溶液转移到另一个试管,在 45 ℃水浴中用氮气吹干。将斑马鱼样本及水样用 1 mL 甲醇溶解,吸进 1 mL 注射器,通过 0.2 μm 有机膜(GHP ACRODISC 13)过滤,制备溶液。通过高效液相色谱-串联质谱仪(ACQUITY UPLC Xevo TQ,Waters)进行液相色谱/质谱联用技术对过滤后的斑马鱼溶液和水溶液中的 PFOA 含量进行测定。液相色谱参考条件:流动相为 5 mmol/L 乙酸铵和甲醇,梯度流速为 200 μL/min,对 PFOA 含量进行测定,分析色谱柱(ACQUITY UPLCr HSS T3 1.8 μm),3 000 V 电压。质谱参考条件:电喷雾离子源为 ESI 源负离子模式,多反应监测模式,单位分辨率,-500 V 喷嘴电压,45 Pa 雾化气压力,260 ℃鞘气温度,11 L/min 鞘气流速,6 L/min 干燥气流速,$-3 500$ V 毛细管电压。

生物富集系数(BCF)是用于评估污染物对环境和人类健康风险的关键指标之一,可用于度量污染物在生物体内的累积趋势。描述鱼类中化学物质生物富集系数的方程为[2]:

$$BCF = \frac{C_f}{C_w} \quad (5-1)$$

式中,C_f 和 C_w 分别代表斑马鱼和溶液中的 PFOA 浓度。

5.2.5 组织病理学分析

为了评估 96 h 暴露情况下微塑料存在/不存在时 PFOA 对斑马鱼不同组织的影响差异,选择 40 mg/L 和 80 mg/L 浓度的 PFOA、40 mg/L 的 PS(0.2 PS、2 PS 和 20 PS)及其混合物进行实验,浓度设置详见表 5-5。暴露期间,控制温度为 (24 ± 1) ℃,光暗周期为 14 h 光照:10 h 黑暗。96 h 暴露后,将来自 12 个暴露组中斑马鱼的鳃、肠和肌肉组织立即放置于 4% PFA 固定液中固定一周。利用热石蜡包埋机(HistoCore Arcadia H Leica ASP)将固定后的组织包埋在石蜡中,利用全自动轮转式切片机(HistocCore NANOCUT R)获得 3 μm 的切片,经封闭式组织脱水机(Leica ASP300S)脱水后用多功能染色机(Leica ST5020-CV5030)对组织进行 H&E 染色,并通过病理切片扫描仪(Pannoramic MIDI)进行观察。

表 5-5 PS 存在/不存在时 PFOA 对斑马鱼组织病理学分析实验浓度

暴露组	对照组	PFOA	0.2 PS	2 PS	20 PS	PFOA+0.2 PS	PFOA+2 PS	PFOA+20 PS
浓度 /(mg/L)	0	40	40	40	40	40+40	40+40	40+40
		80				80+40	80+40	80+40

5.2.6 统计学分析

实验数据用平均值±标准差(SD)表示。通过单因素方差分析(ANOVA)评估测试组和对照组之间的统计学差异。所有实验数据均采用 IBM SPSS 26.0 软件进行统计分析,PFOA、PS(0.2 PS、2 PS 和 20 PS)及其混合物对斑马鱼的 96 h LC_{50} 通过 Probit 分析得到。选择 $P<0.05$ 的统计显著性水平,并对实验结果进行显著性分析。

5.3 PFOA 与 PS-MPs 的急性毒性

PFOA 和 PS-MPs(0.2 PS、2 PS 和 20 PS)单独暴露 96 h 后斑马鱼的存活率曲线如图 5-1 所示,所有空白对照中均未观察到死亡情况。

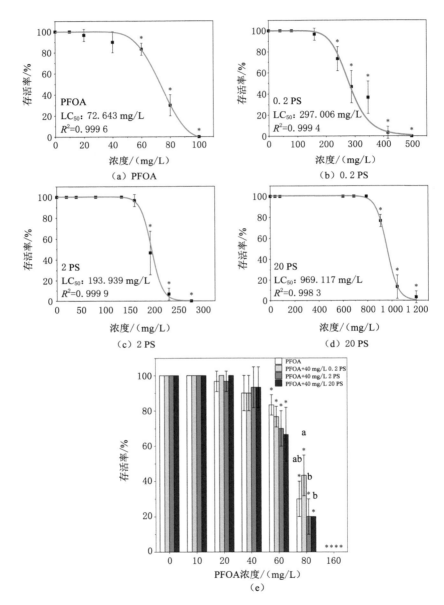

"*"表示与对照相比差异显著($P<0.05$);通过 Waller-Duncan 分析 PFOA、PFOA + 0.2 PS、PFOA + 0.2 PS 和 PFOA + 20 PS 之间差异显著性并在图中用 a、b 表示($P<0.05$)。

图 5-1　斑马鱼暴露于 PFOA、0.2 PS、2 PS 和 20 PS 96 h 的浓度-存活率曲线及浓度-存活率柱状图

第5章 聚苯乙烯微塑料对全氟辛酸斑马鱼毒性效应影响研究

所有暴露条件下,斑马鱼存活率随着污染物浓度的增加而降低。PFOA 的 96 h LC_{50} 为 72.643 mg/L,斑马鱼暴露于 0.2 PS、2 PS 和 20 PS 的 96 h LC_{50} 值分别为 297.006 mg/L、193.939 mg/L 和 969.117 mg/L(表 5-6)。对急性毒性实验结果进行显著性分析发现,与对照组相比,当 PFOA 浓度高于 60 mg/L,0.2 PS、2 PS 和 20 PS 的浓度分别高于 240 mg/L、192 mg/L 和 909 mg/L 时,斑马鱼的存活率开始表现出显著差异($P<0.05$)。急性毒性实验结果表明,三种 PS 对斑马鱼的毒性由高到低依次为 2 PS、0.2 PS 和 20 PS,且 0.2 PS 和 2 PS 对斑马鱼的毒性远高于 20 PS。不同粒径微塑料的生物毒性差异已被很多文献证明,Jeong 等[3]研究发现 0.05 μm 和 0.5 μm PS 显著降低了轮虫的寿命,但 6 μm PS 未造成显著影响。Sun 等[4]发现与 0.7 μm 和 7 μm 相比,0.07 μm PS 对轮虫(*Brachionus plicatilis*)的生存时间、生长和繁殖产生了更显著的负面影响。同时,Lee 等[5]发现 0.05 μm PS 比 0.5 μm 和 6 μm PS 对桡足类(*Tigriopus japonicus*)的生存和繁殖产生的负面影响更大。

表 5-6 PFOA 和 PS 对斑马鱼 96 小时 LC_{50} 和 95% 置信区间($P<0.05$)

暴露组	LC_{50}/(mg/L)	95% 置信区间/(mg/L)
0.2 PS only	297.006	278.730~315.240
2 PS only	193.939	185.883~202.489
20 PS only	969.117	933.683~1 006.898
PFOA only	72.643	65.811~82.511
variable PFOA+40 mg/L 0.2 PS	75.771	68.592~86.944
variable PFOA+40 mg/L 2 PS	66.723	61.055~73.772
variable PFOA+40 mg/L 20 PS	64.520	59.431~70.244

为研究微塑料存在/不存在时 PFOA 对斑马鱼的急性毒性影响,PFOA 和 PS(PFOA+40 mg/L PS)对斑马鱼的毒性实验结果如图 5-1(e)所示。PFOA 分别与 40 mg/L 的 0.2 PS、2 PS 和 20 PS 共同暴露时 96 h LC_{50} 依次为 75.771 mg/L、66.723 mg/L 和 64.520 mg/L。微塑料存在/不存在情况下 PFOA 对斑马鱼存活影响结果显示,96 h 暴露后对斑马鱼 LC_{50} 从大到小依次为 PFOA+20 PS、PFOA+2 PS、PFOA 和 PFOA+0.2 PS。不同暴露情况下,斑马鱼存活率随 PFOA 浓度的升高而不断降低。当 PFOA 浓度从 60 mg/L 开始,暴露组与对照组相比呈现出显著差异($P<0.05$)。当 PFOA 浓度为 80 mg/L PFOA 时,经 Waller-Duncan 分析,发现不同暴露组间存在显著差异,特别是 PFOA+0.2 PS 暴露组与 PFOA+2 PS 暴露组或 PFOA+20 PS 暴露组间差异显著($P<0.05$)。

总之，PS 的存在会影响 PFOA 对斑马鱼存活的毒性。其中，PFOA+20 PS 在各暴露组中毒性最大。微塑料的存在会影响 PFOA 对斑马鱼的毒性，之前的研究得到了类似的结果。Yoo 等[6]发现 48 h 暴露期后，1 mg/L 的 0.05 μm 和 0.5 μm PS 微珠增加了 $HgCl_2$ 对大型蚤 (Diaphanosoma celebensis)的毒性，6 μm PS 没有引起任何影响；0.05 μm PS 增加了 MeHgCl 的毒性，6 μm PS 在减轻了 MeHgCl 的毒性。Tang 等[7]发现苯并[α]芘或 17β-雌二醇与小颗粒 PS(500 nm)共同暴露时，对成年血蛤的免疫毒性增强，当与大颗粒 PS(30 μm)共存时，其对成年血蛤的免疫毒性降低。Hao 等[8]发现 0.6~1.0 μm PS 和敌草隆对淡水硅藻(Cyclotella meneghiniana)和海洋硅藻(Skeletonema costatum)的综合毒性具有拮抗作用。

5.4 PFOA 含量测定及富集分析

表 5-7 显示了所有暴露组中斑马鱼体内和溶液中 PFOA 浓度及生物富集系数(BCF)。暴露 96 h 后所有暴露组斑马鱼体内和溶液中都检测到了 PFOA。对于溶液，微塑料存在与否会影响 PFOA 暴露后残留在溶液中的浓度。除 80 mg/L PFOA 溶液暴露后最终浓度大于初始浓度外，其余各组溶液最终浓度均小于初始浓度。当 PFOA 暴露浓度一定时，存在/不存在微塑料暴露组间暴露后溶液中的 PFOA 浓度差异不大；当 PFOA 暴露浓度高于 20 mg/L 时，不同暴露组间 PFOA+2 PS 暴露组中暴露后溶液中 PFOA 浓度均较低。不同暴露情况下，斑马鱼体内 PFOA 浓度存在差异。PFOA 单独暴露后，斑马鱼体内 PFOA 含量在 PFOA 暴露浓度为 10~80 mg/L 范围内增加缓慢，而在 40~80 mg/L 范围内增加较快。当微塑料存在时，PFOA+2 PS 暴露后，斑马鱼体内 PFOA 含量在 PFOA 暴露浓度为 10~60 mg/L 范围内迅速增加，在 60 mg/L 后增加幅度减小；PFOA+0.2 PS 与 PFOA+20 PS 暴露时，斑马鱼体内 PFOA 含量在 PFOA 暴露浓度为 10~40 mg/L 范围内迅速增加，在 40~60 mg/L 时下降而后又升高[图 5-2(a)]。所有暴露组都观察到 PFOA 在斑马鱼体内的累积。斑马鱼体内 PFOA 浓度与 PFOA 暴露溶液浓度有关。先前的研究发现，(31.8 ± 5.0) μm PS 对 PFOA 有很大的吸附能力，因为其正 ζ 电位对带负电的 PFOA 有很强的静电吸引力[9]。不同大小的微塑料在斑马鱼体内的积累存在差异，Lu 等[10]发现，5 个月大的斑马鱼的鳃、肝和肠中积累了 5 μm PS，而 20 μm PS 仅在鳃和肠中积聚。微塑料的大小影响它们进入斑马鱼体内，也可能影响吸附在微塑料上的污染物进入生物体。

第 5 章 聚苯乙烯微塑料对全氟辛酸斑马鱼毒性效应影响研究

表 5-7 暴露 96 h 后斑马鱼及其水溶液中 PFOA 浓度和生物富集系数

暴露组	PFOA 浓度 /(mg/L)	斑马鱼体内 PFOA 浓度/(μg/g)	溶液中 PFOA 浓度/(μg/mL)	生物富集系数 BCF/(mL/g)
PFOA only	10	64.2	8.3	7.739
	20	86.7	13.6	6.381
	40	159.9	37.6	4.248
	60	273.1	51.5	5.300
	80	396.6	85.3	4.648
variable PFOA+40 mg/L 0.2 PS	10	40.6	9.2	4.391
	20	91.0	13.8	6.583
	40	218.2	32.3	6.762
	60	216.5	50.7	4.266
	80	248.0	78.4	3.163
variable PFOA+40 mg/L 2 PS	10	45.1	8.5	5.307
	20	93.5	13.2	7.075
	40	191.4	31.0	6.166
	60	293.7	48.5	6.051
	80	314.8	74.7	4.215
variable PFOA+40 mg/L 20 PS	10	51.1	8.4	6.114
	20	82.8	13.9	5.947
	40	195.6	32.0	6.113
	60	136.0	54.5	2.496
	80	221.2	75.4	2.935

PFOA 在斑马鱼体内富集情况如图 5-2(b)所示,当 PFOA 单独暴露时,斑马鱼 PFOA 的 BCF 范围约为 4.2～7.8 mL/g,当 PFOA 和 PS 共同暴露时,PFOA 的 BCF 范围约为 2.5～6.8 mL/g。对于 PFOA 暴露组,BCF 随 PFOA 浓度的增加有下降趋势,且在 PFOA 浓度为 40 mg/L 时最低。当微塑料存在时,PFOA+0.2 PS 和 PFOA+2 PS 暴露组中 BCF 呈先升高后降低的趋势。其中,PFOA+0.2 PS 暴露组在 PFOA 浓度为 40 mg/L 时 BCF 最大,PFOA+2 PS 暴露组在 PFOA 浓度为 20 mg/L 时 BCF 最大。PFOA+20 PS 中 PFOA 的 BCF 在 PFOA 浓度低于 40 mg/L 前略有变化,在 PFOA 浓度为 60 mg/L 时骤降,在 80 mg/L 时略有上升。PS 的存在影响了斑马鱼对 PFOA 的 BCF,且斑马鱼对

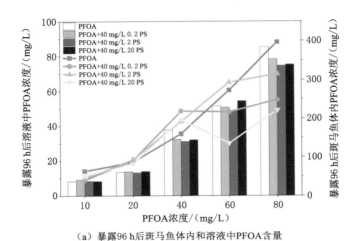

（a）暴露96 h后斑马鱼体内和溶液中PFOA含量

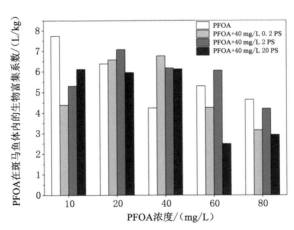

（b）暴露96 h后PFOA在斑马鱼体内生物富集系数

图 5-2　暴露 96 h 后斑马鱼体内和溶液中 PFOA 含量及 PFOA 在斑马鱼体内生物富集系数

PFOA 的 BCF 无明显规律,这可能与暴露实验时间短、PFOA 在斑马鱼体内未达到稳定状态有关。海龟血清中 PFOA 的 BCF 随其在地表水中含量的增加而降低,表明肠道对 PFOA 的吸收可能是一个饱和过程[11]。10~1 000 μg/L 的 PFAS 大部分在 10 d 内在北方豹蛙(Rana pipiens)中达到对的稳定浓度,40 d 内 PFOS 在豹蛙全身的 BCF 为 19.6~119.3 mg/L,而 FPOA 未达到生物浓缩(BCF<1)[11]。此外,Li 等[12]发现 PFOA 能够在食物网中生物累积和营养转移。PFOA 在生物体内的富集可能引起潜在的健康风险,研究人员可以通过扩大 PFOA 浓度范围来观察不同浓度 PFOA 在不同生物体内的富集情况。

5.5 PFOA 与 PS-MPs 对斑马鱼组织的影响

通过 H&E 染色,观察 PFOA、PS 及其混合物暴露后对斑马鱼鳃、肠和肌肉组织的影响。与空白对照组相比,PFOA、PS 及其混合物均未对斑马鱼鳃组织造成明显的病理影响(图 5-3)。Wang 等[13]通过将海洋青鳉暴露于 2 μg/L、20 μg/L 和 200 μg/L 10 μm PS 进行 60 d 慢性实验,观察到海洋青鳉鳃结构受到影响,且微塑料浓度越高鳃受到的损伤越明显。此外,Pei 等[14]发现斑马鱼暴露于 PS 14 d 后鳃受到损伤,且 1 000 mg/L PS 造成的损伤明显高于 100 mg/L,100 nm PS 对鳃的影响高于 50 μm PS。本研究中 PFOA 和 PS 暴露未对斑马鱼鳃造成明显损伤可能与其浓度、暴露时间和微塑料尺寸等因素有关。

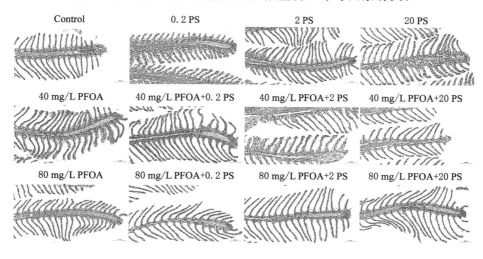

所有暴露组中 PS 浓度均为 40 mg/L。

图 5-3 H&E 染色——96 h 暴露后斑马鱼鳃组织情况

图 5-4 显示斑马鱼暴露于 PFOA、PS 及其混合物 96 h 后其肠组织情况。对于肠组织而言，其受到的影响与污染物浓度有关，浓度越高的 PFOA 对肠的损伤越大，40 mg/L PFOA 对斑马鱼肠基本不造成影响，80 mg/L PFOA 对斑马鱼肠的影响较大。80 mg/L PFOA 对斑马鱼的固有层、柱状上皮细胞和杯状细胞造成损伤。与此类似，Yan 等[15]发现高浓度三氯卡班对斑马鱼肠的损伤高于低浓度三氯卡班。2 PS、0.2 PS 和 20 PS 对斑马鱼肠细胞造成的损伤不明显，2 PS 会引起斑马鱼柱状上皮细胞和杯状细胞受损。对于单独 PS 对斑马鱼肠造成的影响，0.2 PS 和 20 PS 暴露对肠的影响较小，而 2 PS 暴露导致肠柱状上皮受损，杯状细胞分泌较多，这与 2 PS 对斑马鱼急性毒性较大的结果一致。Lu 等[10]发现 5 μm 和 20 μm PS 能进入 5 个月龄斑马鱼肠和鳃，5 μm PS 还能进入斑马鱼肝脏，2 000 μg/L 的 70 nm 和 5 μm PS 在暴露 3 周后均能导致斑马鱼肝脏炎症和脂质堆积。与单一污染物相比，PFOA 和 PS 混合后对斑马鱼肠的影响不同。当 PFOA 浓度为 40 mg/L 时，PFOA+20 PS 暴露造成部分柱状上皮细胞和杯状细胞受损，PFOA+0.2 PS 暴露引起杯状细胞分泌增多。当 PFOA 浓度为 80 mg/L 时，PFOA+20 PS 暴露引起的损伤最严重，严重损害固有层、柱状上皮细胞和杯状细胞。当 PFOA 浓度为 80 mg/L 时，与 0.2 PS 和 2 PS 共存可减轻 PFOA 对斑马鱼肠的损伤，与 20 PS 共存可加强其损伤。结果表明，0.2 PS 或 2 PS 能在一定程度上降低 PFOA 对斑马鱼肠造成的损伤，同时 20 PS 增加了 PFOA 造成的损伤。Zhang 等[16]发现 0.1 μm PS 可降低罗红霉素对红罗非鱼的毒性。Lu 等[17]发现镉和 5 μm PS 微球共存时对斑马鱼肠的影响比镉单独暴露时更严重。与此类似，Paul-Pont 等[18]研究表明 PS 和荧蒽的混合物对贻贝的组织病理损伤比单一污染物更为显著。

图 5-5 显示斑马鱼暴露于 PFOA、PS 及其混合物 96 h 后其肌肉组织情况。对于肌肉组织而言，部分暴露会导致肌纤维裂解，纹理不清晰。80 mg/L PFOA 暴露后对肌肉造成的影响高于低浓度 PFOA。单独微塑料存在时，2 PS 对肌肉的影响较大。微塑料存在时，80 mg/L PFOA+2 PS 暴露后肌肉溶解高于其他暴露组。Fernandes 等[19]发现石墨烯会引起斑马鱼肌肉纤维间炎症浸润。污染物浓度越高，对肌肉组织造成的影响越明显，Hsieh 等[20]研究表明 PE 会损伤太平洋白对虾(*Litopenaeus vannamei*)的肌肉，其损伤与浓度有关，污染物浓度越高影响越明显。PFOA 和 PS 混合物对肌肉组织的影响普遍高于单一污染物，尤其是 PFOA+2 PS 暴露对肌肉纤维溶解的影响。Yang 等[21]研究表明雄性大鼠暴露于二氧化硅纳米颗粒或甲基汞对心脏组织无明显影响，但两者混合后大鼠心肌间隙扩张、肌原纤维紊乱和线粒体损伤明显。简言之，PFOA 浓度越高，对斑马鱼的病理作用越明显。当 PFOA 与不同粒径 PS 共存时，其毒性受到影

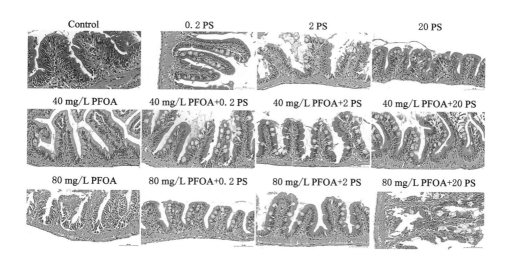

所有暴露组中 PS 浓度均为 40 mg/L。

图 5-4　H&E 染色——96 h 暴露后斑马鱼肠组织情况

响,不同尺寸微塑料能够降低或加强 PFOA 对斑马鱼的影响,具体机制有待进一步研究。

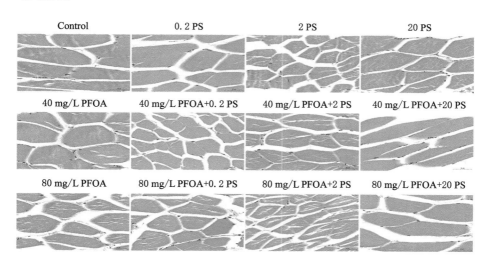

所有暴露组中 PS 浓度均为 40 mg/L。

图 5-5　H&E 染色——96 h 暴露后斑马鱼肌肉组织情况

5.6 本章小结

为研究微塑料对 PFOA 的斑马鱼毒性影响，本章采用典型模式水生生物斑马鱼，选择 PFOA 和粒径为 0.2 μm、2 μm 和 20 μm 的单分散聚苯乙烯微球（0.2 PS、2 PS 和 20 PS）进行暴露实验。通过急性暴露实验、PFOA 含量测定和病理学组织分析研究 PS 存在/不存在时 PFOA 对斑马鱼的毒性影响差异。得出了以下的结论：

① 急性毒性实验结果表明，96 h 暴露后，PFOA 对斑马鱼 LC_{50} 为 72.643 mg/L；0.2 PS、2 PS 和 20 PS 的 96 h LC_{50} 值分别为 297.006 mg/L、193.939 mg/L 和 969.117 mg/L；PFOA 分别与 40 mg/L 的 0.2 PS、2 PS 和 20 PS 共同暴露时 96 h LC_{50} 依次为 75.771 mg/L、66.723 mg/L 和 64.520 mg/L。

② 分析 PFOA 在斑马鱼体内富集情况发现，当 PFOA 单独暴露时，斑马鱼 PFOA 的 BCF 范围约为 4.2~7.8 mL/g；当 PFOA 和 PS 共同暴露时，PFOA 的 BCF 范围约为 2.5~6.8 mL/g。

③ 组织病理学分析结果显示，PS 存在/不存在时 PFOA 均未对斑马鱼鳃组织造成明显影响；当 PFOA 单独存在时，PFOA 浓度越高会对肠和肌肉组织造成更明显的影响；当高浓度 PFOA 与 20 PS 共同暴露时对斑马鱼肠组织造成的影响最明显，当 PFOA 与 2 PS 共同暴露时对斑马鱼肌肉组织造成的损伤更明显。

参考文献

[1] OECD. Test No. 203: fish, acute toxicity test [M/OL]. https://max.book118.com/html/2017/0407/99228612.shtm.

[2] OLIVER B G, NIIMI A J. Bioconcentration factors of some halogenated organics for rainbow trout: limitations in their use for prediction of environmental residues[J]. Environmental science and technology, 1985, 19(9): 842-849.

[3] JEONG C B, WON E J, KANG H M, et al. Microplastic size-dependent toxicity, oxidative stress induction, and p-JNK and p-p38 activation in the monogonont rotifer (*Brachionus koreanus*)[J]. Environmental science and technology, 2016, 50(16): 8849-8857.

[4] SUN Y F, XU W J, GU Q J, et al. Small-sized microplastics negatively

affect rotifers: changes in the key life-history traits and rotifer-Phaeocystis population dynamics[J]. Environmental science and technology,2019,53(15):9241-9251.

[5] LEE K W,SHIM W J,KWON O Y,et al. Size-dependent effects of micro polystyrene particles in the marine copepod *Tigriopus japonicus*[J]. Environmental science and technology,2013,47(19):11278-11283.

[6] YOO J W,JEON M,LEE K W,et al. The single and combined effects of mercury and polystyrene plastic beads on antioxidant-related systems in the brackish water flea: toxicological interaction depending on mercury species and plastic bead size[J]. Aquatic toxicology,2022,252:106325.

[7] TANG Y,RONG J H,GUAN X F,et al. Immunotoxicity of microplastics and two persistent organic pollutants alone or in combination to a bivalve species[J]. Environmental pollution,2020,258:113845.

[8] HAO B B,WU H P,ZHANG S Y,et al. Individual and combined toxicity of microplastics and diuron differs between freshwater and marine diatoms[J]. Science of the total environment,2022,853:158334.

[9] MENG L Y,TIAN H T,LV J T,et al. Influence of microplastics on the photodegradation of perfluorooctane sulfonamide (FOSA)[J]. Journal of environmental sciences,2023,127:791-798.

[10] LU Y F,ZHANG Y,DENG Y F,et al. Uptake and accumulation of polystyrene microplastics in zebrafish (*Danio rerio*) and toxic effects in liver[J]. Environmental science and technology,2016,50(7):4054-4060.

[11] MORIKAWA A,KAMEI N Y,HARADA K,et al. The bioconcentration factor of perfluorooctane sulfonate is significantly larger than that of perfluorooctanoate in wild turtles (*Trachemys scripta elegans* and *Chinemys Reevesii*): an Ai river ecological study in Japan[J]. Ecotoxicology and environmental safety,2006,65(1):14-21.

[12] LI Y N,YAO J Z,ZHANG J,et al. First report on the bioaccumulation and trophic transfer of perfluoroalkyl ether carboxylic acids in estuarine food web[J]. Environmental science and technology,2022,56(10):6046-6055.

[13] WANG J,LI Y J,LU L,et al. Polystyrene microplastics cause tissue damages,sex-specific reproductive disruption and transgenerational effects in marine medaka (*Oryzias melastigma*)[J]. Environmental pollution,2019,254(B):113024.

[14] PEI X, HENG X, CHU W H. Polystyrene nano/microplastics induce microbiota dysbiosis, oxidative damage, and innate immune disruption in zebrafish[J]. Microbial pathogenesis, 2022, 163: 105387.

[15] YAN Z G, DU J Z, ZHANG T X, et al. Impairment of the gut health in Danio rerio exposed to triclocarban[J]. Science of the total environment, 2022, 832: 155025.

[16] ZHANG S S, DING J N, RAZANAJATOVO R M, et al. Interactive effects of polystyrene microplastics and roxithromycin on bioaccumulation and biochemical status in the freshwater fish red tilapia (*Oreochromis niloticus*)[J]. Science of the total environment, 2019, 648: 1431-1439.

[17] LU K, QIAO R X, AN H, et al. Influence of microplastics on the accumulation and chronic toxic effects of cadmium in zebrafish (*Danio rerio*)[J]. Chemosphere, 2018, 202: 514-520.

[18] PAUL-PONT I, LACROIX C, GONZÁLEZ FERNÁNDEZ C, et al. Exposure of marine mussels *Mytilus* spp. to polystyrene microplastics: Toxicity and influence on fluoranthene bioaccumulation[J]. Environmental pollution, 2016, 216: 724-737.

[19] FERNANDES A L, NASCIMENTO J P, SANTOS A P, et al. Assessment of the effects of graphene exposure in *Danio rerio*: a molecular, biochemical and histological approach to investigating mechanisms of toxicity[J]. Chemosphere, 2018, 210: 458-466.

[20] HSIEH S L, WU Y C, XU R Q, et al. Effect of polyethylene microplastics on oxidative stress and histopathology damages in *Litopenaeus vannamei*[J]. Environmental pollution, 2021, 288: 117800.

[21] YANG X Z, FENG L, ZHANG Y N, et al. Co-exposure of silica nanoparticles and methylmercury induced cardiac toxicity *in vitro* and *in vivo*[J]. Science of the total environment, 2018, 631/632: 811-821.

第6章 聚苯乙烯微塑料对全氟辛酸斑马鱼毒性影响机制

6.1 引言

基因组学技术能够揭示个体细胞过程的分子机制,因此在研究环境毒性物质对生物的影响方面得到广泛应用[1]。本章利用基因组学工具,在探究 PFOA 对斑马鱼毒性机制的基础上,进一步研究了不同粒径的 PS-MPs 存在/不存在条件下 PFOA 对斑马鱼毒性机制的影响。

6.2 实验材料与方法

6.2.1 受试生物

受试生物同 5.2.1。

6.2.2 试剂与仪器

PFOA($C_8HF_{15}O_2$,CAS:335-67-1)购自上海阿拉丁生化科技股份有限公司,PFOA 纯度>96%。浓度为 2.5%(v/v)三种不同尺寸的单分散聚苯乙烯微球(PE-MPs,直径分别为 0.2 μm、2 μm 和 20 μm,记作 0.2 PS、2 PS 和 20 PS)购自倍思乐色谱技术开发中心,TRIzol® 试剂、Qubit RNA 检测试剂盒和 Qubit DNA HS 检测试剂盒购自 Invitrogen 公司。

本章所用实验仪器见表 6-1。

表 6-1 实验仪器

仪器名称	仪器型号	生产厂家
超低温冰箱	DW-86L100	青岛海尔生物医疗股份有限公司
超纯水系统	OKP-10	上海涞科仪器有限公司
电子天平	FA2204N	上海菁海仪器有限公司
洁净工作台	DL-CJ-2ND	北京东联哈尔仪器制造有限公司
水质检测仪	HI9829-04	Hanna Instruments
智能人工气候箱	RXZ-1500c-2	宁波江南仪器厂
高效液相色谱-串联质谱仪	ACQUITY UPLC Xevo TQ	Waters
生物分析仪	2100	Agilent

6.2.3 转录组测序

为评估斑马鱼对 PFOA 的分子响应机制,选择 5 个暴露组进行转录组测序分析,分别为对照、PFOA、PFOA+0.2 PS、PFOA+2 PS 和 PFOA+20 PS(PFOA 浓度均为 80 mg/L,PS 浓度均为 40 mg/L)。随机选取 10 条健康斑马鱼放入各暴露组溶液中,进行 3 次生物学重复。实验过程中不进行喂食或换水,控制温度为(24±1) ℃,光暗周期为 14 h 光照∶10 h 黑暗。暴露 96 h 后,使用 TRIzol® 试剂提取斑马鱼的总 RNA。然后使用 Qubit RNA 检测试剂盒对合格的总 RNA 进行定量,并通过安捷伦 2100 生物分析仪(安捷伦技术公司)进行测试。之后使用安捷伦 2100 生物分析仪和 Qubit DNA HS 检测试剂盒对构建的文库进行质量控制和定量,然后在 Illumina Novaseq 6000 测序平台上测序。

6.2.4 生物信息学分析

分别将原对照组和 PFOA 组的结果作为对照组,利用 DESeq2 软件包进行差异表达基因分析。当 $|\log_2 \text{FoldChange}| \geqslant 1$ 且 $P<0.05$ 时该基因被视为差异表达基因。当 $\log_2 \text{FoldChange} \geqslant 1$ 时认为基因上调,当 $\log_2 \text{FoldChange} \leqslant -1$ 时认为基因下调。通过 GO 数据库(http://www.geneontology.org/)对筛选出的差异表达基因进行基因本体功能显著性富集分析,选择 GO 显著富集项($P<0.05$),以了解每个处理影响的关键基因的功能。使用 Reactome 数据库(http://www.reactome.org/)、KEGG 数据库(京都基因和基因组百科全书)(http://www.genome.jp/kegg/)和 Panther 数据库(http://www.pantherdb.org/)对筛选的差异表达基因进行通路富集分析,对显著富集的通路进行分析($P<0.05$)。

6.3 PFOA 对斑马鱼毒性影响的潜在机理

6.3.1 差异表达基因鉴定结果

以 $|\log_2(\text{FoldChange})| \geqslant 1$ 且 $P < 0.05$ 为阈值进行差异表达基因的筛选。结果显示,PFOA 暴露组中共有 843 个差异表达基因,其中 624 个差异基因显著上调($\log_2 \text{FoldChange} \geqslant 1$),219 个差异基因显著下调($\log_2 \text{FoldChange} \leqslant 1$),如图 6-1 所示。

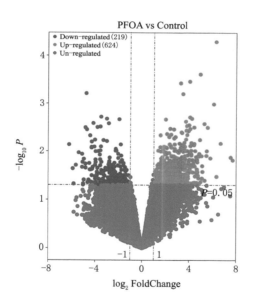

横坐标为差异倍数,以 $\log_2 \text{FoldChange}$ 表示;纵坐标为差异显著性,以 $-\log_{10} P$ 表示。
红色圆点代表上调表达基因,绿色圆点代表下调表达基因,灰色圆点代表没有显著变化的基因。

图 6-1　PFOA 暴露时斑马鱼差异表达基因火山图

6.3.2　PFOA 诱导的斑马鱼差异表达基因的 GO 富集结果

通过分析 PFOA 暴露导致的差异表达基因在 GO 数据库富集情况,进一步了解基因功能,对 PFOA 暴露时差异表达基因进行 GO 富集分析,结果见表 6-2,从生物过程、细胞组分和分子功能三个类别中分别选取显著性最高的十个条目。

表 6-2 PFOA 诱导差异表达基因的 GO 富集结果

GO Term	Database	ID	P 值
ion transport	生物过程	GO:0006811	1.73×10^{-5}
neurotransmitter transport	生物过程	GO:0006836	2.00×10^{-5}
single-organism transport	生物过程	GO:0044765	7.27×10^{-5}
monovalent inorganic cation transport	生物过程	GO:0015672	7.59×10^{-5}
metal ion transport	生物过程	GO:0030001	7.73×10^{-5}
transport	生物过程	GO:0006810	0.000 110 63
cation transport	生物过程	GO:0006812	0.000 176 246
single-organism localization	生物过程	GO:1902578	0.000 192 609
establishment of localization	生物过程	GO:0051234	0.000 207 265
synaptic transmission	生物过程	GO:0007268	0.000 296 388
synapse	细胞组分	GO:0045202	2.49×10^{-6}
synaptic vesicle membrane	细胞组分	GO:0030672	2.74×10^{-5}
transport vesicle membrane	细胞组分	GO:0030658	3.09×10^{-5}
presynapse	细胞组分	GO:0098793	3.13×10^{-5}
synaptic vesicle	细胞组分	GO:0008021	5.85×10^{-5}
neuron part	细胞组分	GO:0097458	6.26×10^{-5}
cytoplasmic vesicle membrane	细胞组分	GO:0030659	0.000 176
vesicle membrane	细胞组分	GO:0012506	0.000 176
plasma membrane	细胞组分	GO:0005886	0.000 186
vesicle	细胞组分	GO:0031982	0.000 193
transporter activity	分子功能	GO:0005215	6.33×10^{-6}
lyase activity	分子功能	GO:0016829	1.43×10^{-5}
ion transmembrane transporter activity	分子功能	GO:0015075	2.84×10^{-5}
transmembrane transporter activity	分子功能	GO:0022857	2.94×10^{-5}
syntaxin binding	分子功能	GO:0019905	0.000 104
metal ion transmembrane transporter activity	分子功能	GO:0046873	0.000 113
substrate-specific transmembrane transporter activity	分子功能	GO:0022891	0.000 12
substrate-specific transporter activity	分子功能	GO:0022892	0.000 358
SNARE binding	分子功能	GO:0000149	0.000 531
inorganic cation transmembrane transporter activity	分子功能	GO:0022890	0.000 567

富集最显著的生物过程条目分别是离子运输(GO:0006811)、神经递质运输(GO:0006836)、单体运输(GO:0044765)、单价无机阳离子运输(GO:0015672)、金属离子运输(GO:0030001)、运输(GO:0006810)、阳离子运输(GO:0006812)、单体定位(GO:1902578)、建立定位(GO:0051234)和突触传递(GO:0007268);富集最显著的细胞组分条目为突触(GO:0045202)、突触囊泡膜(GO:0030672)、运输囊泡膜(GO:0030658)、突触前(GO:0098793)、突触囊泡(GO:0008021)、神经元部分(GO:0097458)、细胞质囊泡膜(GO:0030659)、囊泡膜(GO:0012506)、质膜(GO:0005886)和囊泡(GO:0031982);富集最显著的分子功能条目为转运体活动(GO:0005215)、裂解酶活性(GO:0016829)、离子跨膜转运蛋白活性(GO:0015075)、跨膜转运蛋白活性(GO:0022857)、突触融合蛋白结合(GO:0019905)、金属离子跨膜转运蛋白活性(GO:0046873)、底物特异性跨膜转运蛋白活性(GO:0022891)、底物特异性转运蛋白活性(GO:0022892)、SNARE结合(GO:0000149)和无机阳离子跨膜转运蛋白活性(GO:0022890)。为进一步研究PFOA对生物过程、分子功能和细胞组分的影响,统计分析差异表达基因二级分类情况,结果发现PFOA暴露时斑马鱼的差异表达基因显著富集于细胞过程(上调335,下调88)、单体过程(上调294,下调75)、生物调节(上调221,下调41)、代谢过程(上调168,下调75)等生物过程;作用于结合(上调320,下调76)、催化活性(上调117,下调68)、转运活性(上调57,下调14)等分子功能;影响细胞(上调257,下调64)、细胞部分(上调257,下调64)、膜(上调221,下调54)、膜部分(上调189,下调48)等细胞组分(图6-2)。

6.3.3 PFOA暴露差异表达基因通路富集结果

通过分析PFOA暴露时差异表达基因在KEGG、Reactome和Panther数据库中参与功能通路的情况,进一步了解差异表达基因与生物学通路的关联。基于显著性筛选富集最显著的前15个通路($P<0.05$)(图6-3)。PFOA暴露时最显著的前15个通路为化学突触间的传递(R-DRE-112315)、神经系统(R-DRE-112316)、去甲肾上腺素神经递质释放周期(R-DRE-181430)、GABA A受体激活(R-DRE-977441)、乙酰胆碱神经递质释放周期(R-DRE-264642)、突触后细胞中神经递质受体的结合与下游传递(R-DRE-112314)、谷氨酸神经递质释放周期(R-DRE-210500)、@@GABA受体激活(R-DRE-977443)、配体门控离子通道转运(R-DRE-975298)、离子通道传输(R-DRE-983712)、第0阶段:快速去极化(R-DRE-5576892)、第2阶段:平台阶段(R-DRE-5576893)、心肌收缩(DRE04260)、第1阶段:快速Na^+通道失活(R-DRE-5576894)和γ-氨基丁酸的

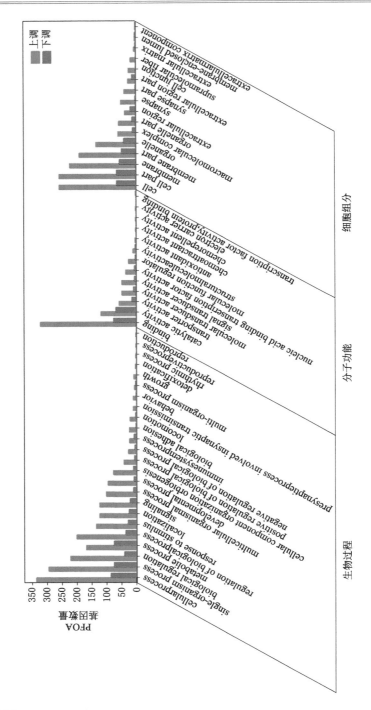

图6-2 PFOA暴露差异表达基因GO富集二级分类统计图

红色部分为上调基因;绿色部分为下调基因;纵坐标为该条目富集的基因数。

合成、释放、再摄取和降解(R-DRE-888590)。

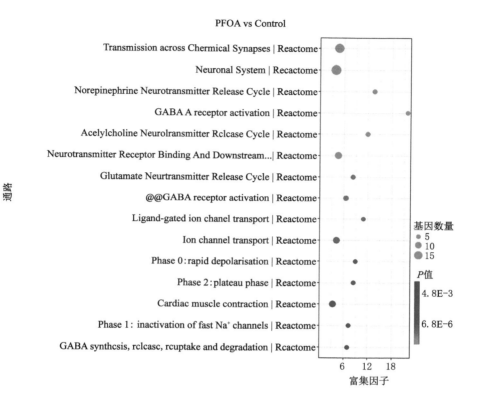

横坐标表示有注释的差异基因占总差异基因的频率与注释的背景基因占总背景基因的频率的比值,纵坐标为富集最显著的前15个通路;颜色代表显著性,颜色越红表示越显著;气泡大小代表富集基因的数量,气泡越大表示富集基因越多。

图6-3 PFOA暴露时差异表达基因通路富集气泡图

6.3.4 PFOA对斑马鱼神经递质释放周期的影响

通路分析结果表明,PFOA显著影响神经系统中跨化学突触的传递,特别是神经递质释放周期,如去甲肾上腺素神经递质释放周期、谷氨酸神经递质释放周期和乙酰胆碱神经递质释放周期(图6-4),GO富集结果可以印证上述观点,PFOA显著影响斑马鱼神经系统中化学突触的传递。化学突触是突触前神经元与其各自的突触后靶点之间信号传导的细胞-细胞接触的特殊位点[2],也是神经元与其伴侣细胞之间的接触和信息传递的位点[3]。化学突触的信号传导过程

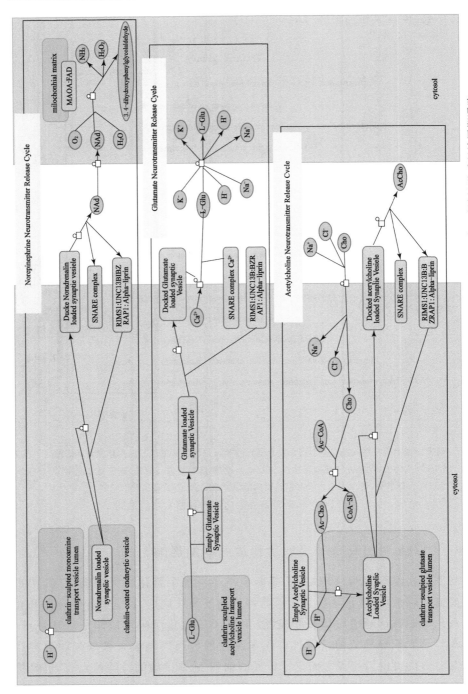

图6-4 去甲肾上腺素神经递质释放周期、谷氨酸神经递质释放周期和乙酰胆碱神经递质释放周期通路

包含充满神经递质的突触囊泡,囊泡与质膜融合以释放神经递质,随后进行重组和回收,体外突触的刺激通常会导致大量突触囊泡释放神经递质以维持高活性传递[4]。斑马鱼作为一种用于神经生物学和神经毒性生态学的脊椎动物模型,与哺乳动物共享多种神经递质,包括谷氨酸、GABA、甘氨酸、多巴胺、去甲肾上腺素、肾上腺素、血清素、乙酰胆碱和组胺[5]。去甲肾上腺素是周围神经系统和中枢神经系统中的神经递质和激素[6],谷氨酸具有兴奋性毒性,在神经元细胞死亡中起关键作用[7],它还是中枢神经系统中最丰富的兴奋性神经递质,通过代谢性谷氨酸受体调节突触传递和神经元兴奋性[8]。乙酰胆碱由胆碱乙酰转移酶合成,由突触前神经末梢分泌,与聚集在突触后膜的乙酰胆碱受体结合。释放后,它被属于 B 型羧酸酯酶家族的乙酰胆碱酯酶从突触间隙迅速代谢,并将乙酰胆碱裂解为胆碱和乙酸盐[9]。在神经元胞吐过程中起关键作用的基因 $snap25b$[10]等显著上调。上述基因均与化学突触的传递中神经递质释放周期有关,因此,PFOA 暴露会通过上调有关基因显著影响斑马鱼体内化学突触传递通路进而对神经系统造成影响。

6.4 PS-MPs 与 PFOA 对斑马鱼的毒性影响的潜在机理

6.4.1 差异表达基因鉴定结果

以 $|\log_2 \text{FoldChange}| \geqslant 1$ 且 $P<0.05$ 为阈值进行差异表达基因的筛选。PFOA+0.2 PS 暴露组中共有 222 个差异表达基因,其中 119 个基因显著上调、103 个基因显著下调;PFOA+2 PS 暴露组中共有 500 个差异表达基因,其中 137 个基因显著上调、363 个基因显著下调;PFOA+20 PS 暴露组中共有 370 个差异表达基因,其中 80 个基因显著上调、290 个基因显著下调,如图 6-5 所示。

6.4.2 PS-MPs 与 PFOA 诱导的斑马鱼差异表达基因的 GO 富集结果

通过分析 PFOA+0.2 PS 暴露导致的差异表达基因在 GO 数据库富集情况,进一步了解基因功能。分析 PFOA+0.2 PS 暴露组,分别从生物过程、细胞组分和分子功能三个类别中各选取富集显著性最高的十个条目。富集最显著的生物过程条目分别是单体运输(GO:0044765)、单体定位(GO:1902578)、神经递质运输(GO:0006836)、运输(GO:0006810)、建立定位(GO:0051234)、离子运输(GO:0006811)、γ-氨基丁酸转运(GO:0015812)、跨膜运输(GO:0055085)、氮化合物运输(GO:0071705)和一元羧酸运输(GO:0015718);富集最显著的细

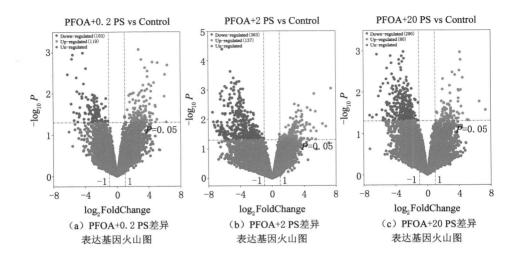

横坐标为差异倍数,以 $\log_2 \text{FoldChange}$ 表示;纵坐标为差异显著性,以 $-\log_{10}P$ 表示。红色圆点代表上调表达基因,绿色圆点代表下调表达基因,灰色圆点代表没有显著变化的基因。

图 6-5　PFOA+0.2 PS、PFOA+2 PS 和 PFOA+20 PS 差异表达基因火山图

胞组分条目为质膜部分(GO:0044459)、质膜的组成部分(GO:0005887)、质膜固有成分(GO:0031226)、质膜(GO:0005886)、细胞外围(GO:0071944)、外动力臂(GO:0036157)、细胞投影(GO:0042995)、顶端质膜(GO:0016324)、细胞顶端(GO:0045177)和神经元投影(GO:0043005);富集最显著的分子功能条目为阴离子-阳离子转运体活性(GO:0015296)、γ-氨基丁酸跨膜转运蛋白活性(GO:0015185)、γ-氨基丁酸:钠转运体活性(GO:0005332)、钠-氨基酸转运体活性(GO:0005283)、离子跨膜转运蛋白活性(GO:0015075)、转运活性(GO:0005215)、阳离子-氨基酸转运体活性(GO:0005416)、底物特异性转运蛋白活性(GO:0008324)、金属离子跨膜转运蛋白活性(GO:0046873)和跨膜转运蛋白活性(GO:0022857),见表 6-3。如图 6-6(a)所示,对差异表达基因在 GO 二级分类情况进行分析,PFOA+0.2 PS 暴露时斑马鱼的差异表达基因显著富集于细胞过程(上调 52,下调 36)、单体过程(上调 46,下调 36)、生物调节(上调 38,下调 16)、生物过程调控(上调 36,下调 13)等生物过程;作用于结合(上调 49,下调 26)、催化活性(上调 18,下调 18)、转运活性(上调 4,下调 17)等分子功能;影响膜(上调 35,下调 36)、细胞(上调 39,下调 32)、细胞部分(上调 39,下调 32)、膜部分(上调 30,下调 34)等细胞组分。

第6章 聚苯乙烯微塑料对全氟辛酸斑马鱼毒性影响机制

表 6-3 PFOA+0.2 PS 暴露诱导差异表达基因的 GO 富集结果

GO Term	Database	ID	P 值
single-organism transport	生物过程	GO:0044765	2.27×10^{-5}
single-organism localization	生物过程	GO:1902578	3.88×10^{-5}
neurotransmitter transport	生物过程	GO:0006836	6.85×10^{-5}
transport	生物过程	GO:0006810	9.37×10^{-5}
establishment of localization	生物过程	GO:0051234	0.000 128 803
ion transport	生物过程	GO:0006811	0.000 178 515
gamma-aminobutyric acid transport	生物过程	GO:0015812	0.000 268 943
transmembrane transport	生物过程	GO:0055085	0.000 436 587
nitrogen compound transport	生物过程	GO:0071705	0.000 613 899
monocarboxylic acid transport	生物过程	GO:0015718	0.000 862 111
single-organism transport	细胞组分	GO:0044459	3.95×10^{-5}
single-organism localization	细胞组分	GO:0005887	0.000 12
neurotransmitter transport	细胞组分	GO:0031226	0.000 142
transport	细胞组分	GO:0005886	0.000 144
establishment of localization	细胞组分	GO:0071944	0.000 194
ion transport	细胞组分	GO:0036157	0.006 776
gamma-aminobutyric acid transport	细胞组分	GO:0042995	0.006 998
transmembrane transport	细胞组分	GO:0016324	0.007 941
nitrogen compound transport	细胞组分	GO:0045177	0.009 572
monocarboxylic acid transport	细胞组分	GO:0043005	0.012 234
anion:cation symporter activity	分子功能	GO:0015296	1.53×10^{-6}
gamma-aminobutyric acid transmembrane transporter activity	分子功能	GO:0015185	1.02×10^{-5}
gamma-aminobutyric acid:sodium symporter activity	分子功能	GO:0005332	1.02×10^{-5}
sodium:amino acid symporter activity	分子功能	GO:0005283	1.62×10^{-5}
ion transmembrane transporter activity	分子功能	GO:0015075	2.07×10^{-5}
transporter activity	分子功能	GO:0005215	2.29×10^{-5}
cation:amino acid symporter activity	分子功能	GO:0005416	2.42×10^{-5}
cation transmembrane transporter activity	分子功能	GO:0008324	2.60×10^{-5}
metal ion transmembrane transporter activity	分子功能	GO:0046873	2.67×10^{-5}
transmembrane transporter activity	分子功能	GO:0022857	2.99×10^{-5}

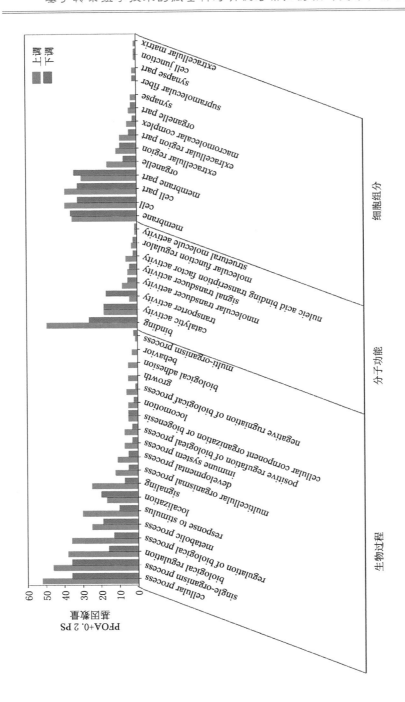

图6-6 PFOA+0.2 PS、PFOA+2 PS和PFOA+20 PS暴露下差异表达基因的GO二级分类统计图

横坐标根据生物过程、分子功能和细胞组分进行分类，纵坐标为该条目富集的基因数，红色部分为上调基因，绿色部分为下调基因。

第6章 聚苯乙烯微塑料对全氟辛酸斑马鱼毒性影响机制

(b) PFOA+2 PS

图6-6 （续）

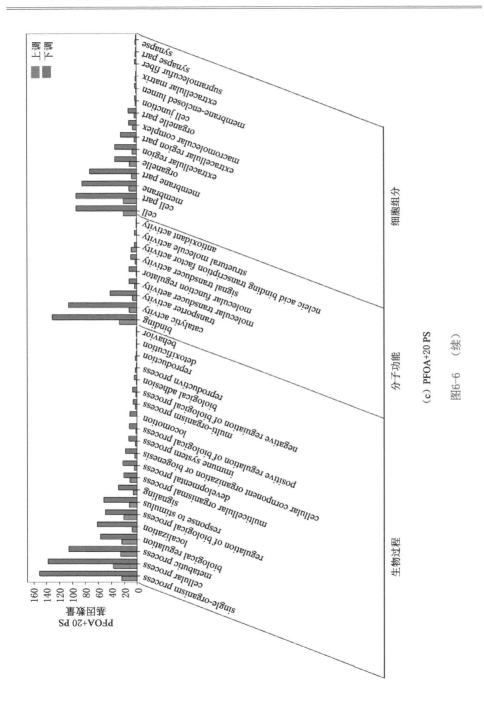

(c) PFOA+20 PS

图6-6 （续）

通过分析PFOA+2 PS暴露导致的差异表达基因在GO数据库富集情况，进一步了解基因功能。分析PFOA+2 PS暴露组，分别从生物过程、细胞组分和分子功能三个类别中各选取富集显著性最高的十个条目。富集最显著的生物过程条目分别是阴离子运输(GO:0006820)、跨膜运输(GO:0055085)、单体运输(GO:0044765)、有机酸运输(GO:0015849)、羧酸运输(GO:0046942)、有机阴离子运输(GO:0015711)、羧酸代谢过程(GO:0019752)、前肾发育中涉及的模式规范(GO:0039017)、前/后模式规范(GO:0034672)和肾系统模式规范(GO:0072048)；富集最显著的细胞组分条目为质膜的组成部分(GO:0005887)、质膜固有成分(GO:0031226)、质膜部分(GO:0044459)、细胞外间隙(GO:0005615)、胞外区部分(GO:0044421)、细胞外区域(GO:0005576)、质膜(GO:0005886)、细胞外围(GO:0071944)、神经元投影(GO:0043005)和P颗粒(GO:0043186)；富集最显著的分子功能条目为次级活性跨膜转运蛋白活性(GO:0015291)、一元羧酸跨膜转运蛋白活性(GO:0008028)、羧酸跨膜转运蛋白活性(GO:0046943)、活性跨膜转运蛋白活性(GO:0022804)、有机酸跨膜转运蛋白活性(GO:0005342)、阴离子跨膜转运蛋白活性(GO:0008509)、同转运体活性(GO:0015293)、有机阴离子跨膜转运蛋白活性(GO:0008514)、溶质-阳离子转运体活性(GO:0015294)和钠离子跨膜转运蛋白活性(GO:0015081)，见表6-4。如图6-6(b)所示，对差异表达基因在GO二级分类情况进行分析，PFOA+2 PS暴露时斑马鱼的差异表达基因显著富集于单体过程(上调47,下调172)、细胞过程(上调52,下调164)、代谢过程(上调31,下调132)、生物调节(上调32,下调70)等生物过程；作用于结合(上调45,下调145)、催化活性(上调20,下调117)、转运活性(上调5,下调54)等分子功能；影响细胞(上调36,下调124)、细胞部分(上调36,下调124)、膜(上调31,下调110)、膜部分(上调28,下调97)等细胞组分。

表6-4 PFOA+2 PS暴露诱导差异表达基因的GO富集分析结果

GO Term	Database	ID	P值
anion transport	生物过程	GO:0006820	1.15×10^{-8}
transmembrane transport	生物过程	GO:0055085	5.65×10^{-7}
single-organism transport	生物过程	GO:0044765	1.80×10^{-6}
organic acid transport	生物过程	GO:0015849	2.14×10^{-6}
carboxylic acid transport	生物过程	GO:0046942	2.14×10^{-6}
organic anion transport	生物过程	GO:0015711	2.56×10^{-6}

表 6-4(续)

GO Term	Database	ID	P 值
carboxylic acid metabolic process	生物过程	GO:0019752	3.39×10^{-6}
pattern specification involved in pronephros development	生物过程	GO:0039017	3.90×10^{-6}
anterior/posterior pattern specification involved in pronephros development	生物过程	GO:0034672	3.90×10^{-6}
renal system pattern specification	生物过程	GO:0072048	3.90×10^{-6}
integral component of plasma membrane	细胞组分	GO:0005887	2.97×10^{-7}
intrinsic component of plasma membrane	细胞组分	GO:0031226	4.14×10^{-7}
plasma membrane part	细胞组分	GO:0044459	8.42×10^{-6}
extracellular space	细胞组分	GO:0005615	5.09×10^{-5}
extracellular region part	细胞组分	GO:0044421	0.000 295
extracellular region	细胞组分	GO:0005576	0.001 032
plasma membrane	细胞组分	GO:0005886	0.003 718
cell periphery	细胞组分	GO:0071944	0.005 158
neuron projection	细胞组分	GO:0043005	0.005 259
P granule	细胞组分	GO:0043186	0.010 328
secondary active transmembrane transporter activity	分子功能	GO:0015291	5.91×10^{-14}
monocarboxylic acid transmembrane transporter activity	分子功能	GO:0008028	7.09×10^{-13}
carboxylic acid transmembrane transporter activity	分子功能	GO:0046943	1.36×10^{-12}
active transmembrane transporter activity	分子功能	GO:0022804	1.96×10^{-12}
organic acid transmembrane transporter activity	分子功能	GO:0005342	1.99×10^{-12}
anion transmembrane transporter activity	分子功能	GO:0008509	2.75×10^{-12}
symporter activity	分子功能	GO:0015293	3.1×10^{-12}
organic anion transmembrane transporter activity	分子功能	GO:0008514	6.93×10^{-12}
solute:cation symporter activity	分子功能	GO:0015294	3.96×10^{-11}
sodium ion transmembrane transporter activity	分子功能	GO:0015081	1.77×10^{-10}

通过分析 PFOA+20 PS 暴露导致的差异表达基因在 GO 数据库富集情况，进一步了解基因功能。分析 PFOA+20 PS 暴露组，分别从生物过程、细胞组分和分子功能三个类别中各选取富集显著性最高的十个条目。富集最显著的生物过程条目分别是羧酸代谢过程(GO:0019752)、含氧酸代谢过程(GO:0043436)、有机酸代谢过程(GO:0006082)、小分子代谢过程(GO:0044281)、一

第6章 聚苯乙烯微塑料对全氟辛酸斑马鱼毒性影响机制

元羧酸代谢过程（GO：0032787）、α-氨基酸代谢过程（GO：1901605）、跨膜运输（GO：0055085）、细胞氨基酸代谢过程（GO：0006520）、细胞生物胺分解代谢过程（GO：0042402）和芳香族氨基酸家族分解代谢过程（GO：0009074）；富集最显著的细胞组分条目为质膜的组成部分（GO：0005887）、细胞外间隙（GO：0005615）、质膜固有成分（GO：0031226）、质膜部分（GO：0044459）、细胞外区域部分（GO：0044421）、细胞外区域（GO：0005576）、细胞外围（GO：0071944）、质膜（GO：0005886）、ATP结合盒（ABC）转运蛋白复合体（GO：0043190）和极低密度脂蛋白颗粒（GO：0034361）；富集最显著的分子功能条目为阴离子跨膜转运蛋白活性（GO：0008509）、活性跨膜转运蛋白活性（GO：0022804）、溶质-阳离子转运体活性（GO：0015294）、钠离子跨膜转运蛋白活性（GO：0015081）、次级活性跨膜转运蛋白活性（GO：0015291）、氧化还原酶活性（GO：0016491）、阴离子-阳离子转运体活性（GO：0015296）、有机阴离子跨膜转运蛋白活性（GO：0008514）、跨膜转运蛋白活性（GO：0022857）和一元羧酸跨膜转运蛋白活性（GO：0008028），见表6-5。如图6-6(c)所示，对差异表达基因在GO二级分类情况进行分析，PFOA+20 PS暴露时斑马鱼的差异表达基因显著富集于生物过程调控单体过程（上调24，下调151）、细胞过程（上调37，下调138）、代谢过程（上调26，下调106）、生物调节（上调24，下调57）等生物过程；作用于分子功能调节剂结合（上调27，下调131）、催化活性（上调12，下调106）、转运活性（上调7，下调42）等分子功能；影响细胞（上调21，下调94）、细胞部分（上调21，下调94）、膜（上调12，下调85）、膜部分（上调8，下调73）等细胞组分。

表6-5 PFOA+20 PS暴露诱导差异表达基因的GO富集分析结果

GO Term	Database	ID	P值
carboxylic acid metabolic process	生物过程	GO：0019752	6.20×10^{-9}
oxoacid metabolic process	生物过程	GO：0043436	1.87×10^{-8}
organic acid metabolic process	生物过程	GO：0006082	1.98×10^{-8}
small molecule metabolic process	生物过程	GO：0044281	3.86×10^{-8}
monocarboxylic acid metabolic process	生物过程	GO：0032787	5.31×10^{-7}
alpha-amino acid metabolic process	生物过程	GO：1901605	8.64×10^{-7}
transmembrane transport	生物过程	GO：0055085	6.54×10^{-6}
cellular amino acid metabolic process	生物过程	GO：0006520	1.92×10^{-5}
cellular biogenic amine catabolic process	生物过程	GO：0042402	2.12×10^{-5}
aromatic amino acid family catabolic process	生物过程	GO：0009074	2.12×10^{-5}

表 6-5(续)

GO Term	Database	ID	P 值
integral component of plasma membrane	细胞组分	GO:0005887	2.53×10^{-5}
extracellular space	细胞组分	GO:0005615	2.64×10^{-5}
intrinsic component of plasma membrane	细胞组分	GO:0031226	3.23×10^{-5}
plasma membrane part	细胞组分	GO:0044459	6.57×10^{-5}
extracellular region part	细胞组分	GO:0044421	0.000 447
extracellular region	细胞组分	GO:0005576	0.000 722
cell periphery	细胞组分	GO:0071944	0.000 980
plasma membrane	细胞组分	GO:0005886	0.001 517
ATP-binding cassette (ABC) transporter complex	细胞组分	GO:0043190	0.013 032
very-low-density lipoprotein particle	细胞组分	GO:0034361	0.013 032
anion transmembrane transporter activity	分子功能	GO:0008509	1.55×10^{-7}
active transmembrane transporter activity	分子功能	GO:0022804	2.94×10^{-7}
solute:cation symporter activity	分子功能	GO:0015294	4.84×10^{-7}
sodium ion transmembrane transporter activity	分子功能	GO:0015081	1.25×10^{-6}
secondary active transmembrane transporter activity	分子功能	GO:0015291	1.30×10^{-6}
oxidoreductase activity	分子功能	GO:0016491	1.96×10^{-6}
anion:cation symporter activity	分子功能	GO:0015296	2.00×10^{-6}
organic anion transmembrane transporter activity	分子功能	GO:0008514	2.15×10^{-6}
transmembrane transporter activity	分子功能	GO:0022857	2.32×10^{-6}
monocarboxylic acid transmembrane transporter activity	分子功能	GO:0008028	2.45×10^{-6}

6.4.3 PS-MPs 与 PFOA 诱导的斑马鱼差异表达基因的通路富集结果

通过分析 PFOA+0.2 PS 暴露导致的差异表达基因在 KEGG、Reactome 和 Panther 通路中的功能参与,进一步了解差异表达基因与生物学通路的关联[图 6-7(a)]。基于显著性筛选富集最显著的前 15 个通路($P<0.05$)。PFOA+0.2 PS 暴露时最显著的前 15 个通路是叶酸和的代谢(R-DRE-196757)、激活 G 蛋白门控钾通道(R-DRE-1296041)、通过 Gbeta/gamma 亚基抑制电压门控 Ca^{2+} 通道(R-DRE-997272)、G 蛋白门控钾通道(R-DRE-1296059)、糖鞘脂生

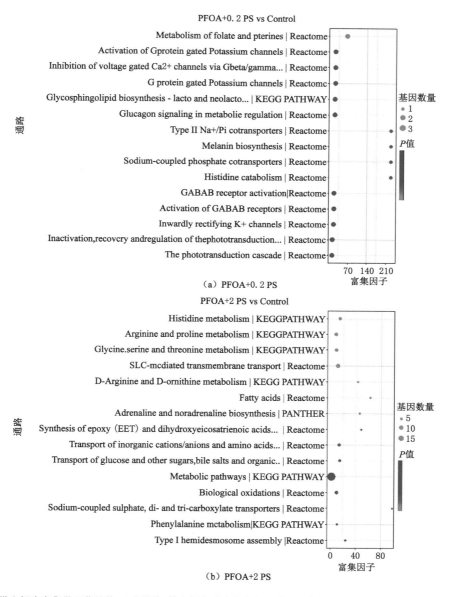

纵坐标为富集最显著的前15个通路,横坐标表示有注释的差异基因占总差异基因的频率与注释的
背景基因占总背景基因的频率的比值;颜色代表显著性,颜色越红表示越显著;
气泡大小代表富集基因的数量,气泡越大表示富集基因越多。

图6-7　PFOA+0.2 PS、PFOA+2 PS 和 PFOA+20 PS暴露时
差异表达基因通路富集气泡图

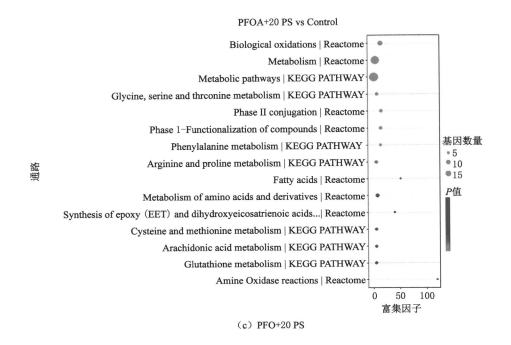

(c) PFO+20 PS

图 6-7 （续）

合成-乳糖和新乳糖系列（DRE00601）、胰高血糖素信号在代谢调节中的作用（R-DRE-163359）、Ⅱ型 Na^+/Pi 共转运蛋白（R-DRE-427589）、黑色素生物合成（R-DRE-5662702）、钠偶联磷酸盐共转运蛋白（R-DRE-427652）、组氨酸分解代谢（R-DRE-70921）、GABA B 受体激活（R-DRE-977444）、GABA B 受体的激活（R-DRE-991365）、内整流 K^+ 通道（R-DRE-1296065）、光转导级联的失活-恢复-调节（R-DRE-2514859）和光转导级联（R-DRE-2514856）。

通过分析 PFOA＋2 PS 暴露时差异表达基因在 KEGG、Reactome 和 Panther 数据库中参与功能通路的情况,进一步了解差异表达基因与生物学通路的关联[图 6-7(b)]。PFOA＋2 PS 暴露时最显著的前 15 个通路是组氨酸代谢（DRE00340）,精氨酸和脯氨酸代谢（DRE00330）,甘氨酸、丝氨酸和苏氨酸代谢（DRE00260）,SLC 介导的跨膜转运（R-DRE-425407）,d-精氨酸和 d-鸟氨酸代谢（DRE00472）,脂肪酸代谢（R-DRE-211935）,肾上腺素和去甲肾上腺素生物合成（P00001）,环氧树脂（EET）和二羟基二十碳三烯酸（DHET）的合成（R-DRE-

2142670),无机阳离子/阴离子和氨基酸/寡肽的转运(R-DRE-425393),葡萄糖和其他糖、胆盐和有机酸、金属离子和胺类化合物的运输(R-DRE-425366),代谢途径(DRE01100),生物氧化反应(R-DRE-211859),钠偶联硫酸盐、二羧酸和三羧酸转运体(R-DRE-433137),苯丙氨酸代谢(DRE00360)和Ⅰ型半桥粒组件(R-DRE-446107)。

通过分析 PFOA+20 PS 暴露时差异表达基因在 KEGG、Reactome 和 Panther 数据库中参与功能通路的情况,进一步了解差异表达基因与生物学通路的关联[图 6-7(c)]。PFOA+20 PS 暴露时最显著的前 15 个通路是生物氧化(R-DRE-211859),新陈代谢(R-DRE-1430728),代谢途径(DRE01100),甘氨酸、丝氨酸和苏氨酸代谢(DRE00260),第Ⅱ阶段共轭(R-DRE-156580),阶段Ⅰ:化合物的功能化(R-DRE-211945),苯丙氨酸代谢(DRE00360),精氨酸和脯氨酸代谢(DRE00330),脂肪酸代谢(R-DRE-211935),氨基酸及其衍生物的代谢(R-DRE-71291),环氧树脂(EET)和二羟基二十碳三烯酸(DHET)的合成(R-DRE-2142670),半胱氨酸和蛋氨酸代谢(DRE00270),花生四烯酸代谢(DRE00590),谷胱甘肽代谢(DRE00480)和胺氧化酶反应(R-DRE-140179)。

6.4.4 PS-MPs 与 PFOA 对斑马鱼内整流 K^+ 通道通路的影响

PFOA+0.2 PS 显著影响神经系统中的钾通道(图 6-8),钾通道是家族和功能特性数量最多样化的通道组[10],包括 Ca^{2+} 激活 K^+ 通道、氢氰酸通道、内整流 K^+ 通道、电压门控通道和串联孔隙结构域的钾离子通道。可以通过调节胆碱能活性来调节认知功能的基因 *gng*8[11]和基因 *gnb*3*a* 显著上调,并富集在神经系统中内整流 K^+ 通道通路。结果显示,PFOA 与 0.2 PS 共同暴露会显著影响内整流 K^+ 通道通路中 K^+ 离子的跨膜运输,进而影响斑马鱼神经系统。通过影响神经递质 γ-氨基丁酸(GABA)、G 蛋白 β-γ 复合物、GABA B 受体 G 蛋白 β-γ 复合物影响 GABA B 受体 G 蛋白和 Kir3 通道复合物影响 K^+ 从膜外到膜内的被动运输,ATP 敏感性 K^+ 通道,ATP 敏感性 K^+ 通道向内整流和钾转运通道(Kir1.1 和 Kir4.1/5.1)影响 K^+ 从膜内到膜外的主动运输。

6.4.5 PS-MPs 与 PFOA 对斑马鱼甘氨酸、丝氨酸和苏氨酸代谢通路的影响

PFOA+2 PS 和 PFOA+20 PS 两种暴露均对甘氨酸、丝氨酸和苏氨酸代谢通路造成显著影响(图 6-9)。甘氨酸具有重要的抗氧化功能,是中枢神经系统主要抑制的神经递质[12]。丝氨酸来源于营养素的代谢,包括蛋白质和磷脂。

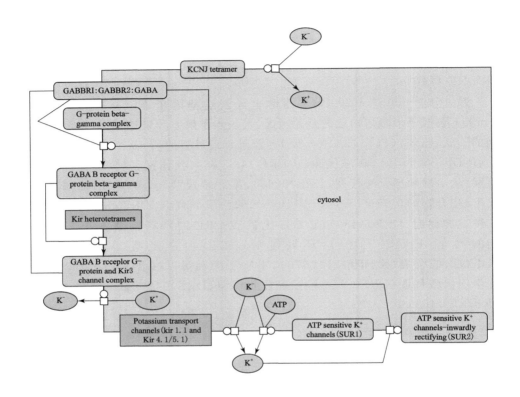

图 6-8 内整流 K^+ 通道通路

丝氨酸可以从甘氨酸内源性合成,甘氨酸通过丝氨酸羟甲基转移酶(SHMT)获得一个羟甲基以产生丝氨酸[13]。苏氨酸是一种必需氨基酸,L-苏氨酸是合成蛋白质所必需的,是甘氨酸的前体[14]。PFOA+2 PS 和 PFOA+20 PS 暴露引起基因 *agxtb*、*dao.2*、*mao* 显著下调。此外,PFOA+2 PS 暴露还导致基因 *grhprb* 和 *gatm* 显著下调,PFOA+20 PS 暴露还导致基因 *bhmt*、*aoc2* 和 *gcshb* 显著下调。PFOA+2 PS 和 PFOA+20 PS 暴露均导致斑马鱼差异表达基因显著下调,对甘氨酸、丝氨酸和苏氨酸代谢通路造成影响,包括丝氨酸的激活以及甘氨酸与丝氨酸间的转化,未影响甘氨酸及苏氨酸的转化。

第6章 聚苯乙烯微塑料对全氟辛酸斑马鱼毒性影响机制

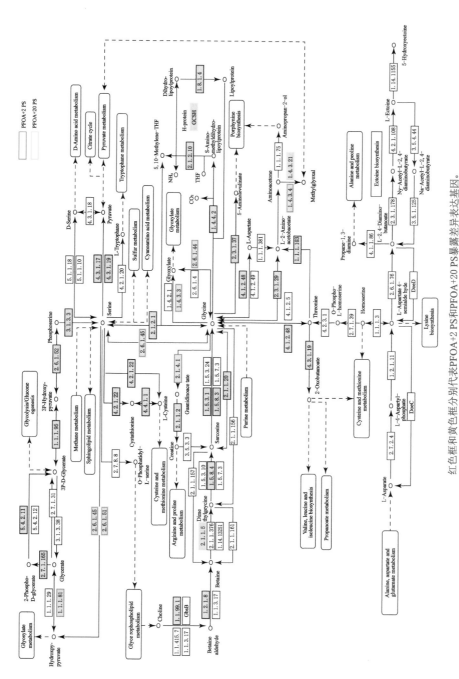

图6-9 甘氨酸、丝氨酸和苏氨酸代谢通路

红色框和黄色框分别代表PFOA+2 PS和PFOA+20 PS暴露差异表达基因。

6.5 不同粒径 PS-MPs 对 PFOA 斑马鱼的毒性影响的差异

6.5.1 PS-MPs 粒径对 PFOA 诱导差异表达基因的影响

以 PFOA 暴露组为对照,以 $|\log_2 \text{FoldChange}| \geqslant 1$ 且 $P < 0.05$ 为阈值进行差异表达基因的筛选。筛选结果如图 6-10 所示,与 PFOA 组相比,PFOA+0.2 PS 暴露组中共有 390 个基因差异表达,其中 139 个基因显著上调、251 个基因显著下调;PFOA+2 PS 暴露组中共有 869 个基因差异表达,其中 274 个基因显著上调、595 个基因显著下调;PFOA+20 PS 暴露组中共有 2 854 个基因差异表达,其中 1 736 个基因显著上调、1 118 个基因显著下调。

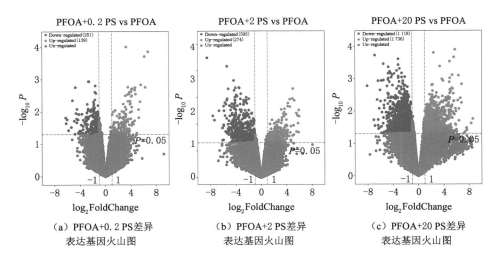

(a) PFOA+0.2 PS 差异表达基因火山图　(b) PFOA+2 PS 差异表达基因火山图　(c) PFOA+20 PS 差异表达基因火山图

横坐标为差异倍数,以 $\log_2 \text{FoldChange}$ 表示;纵坐标为差异显著性,以 $-\log_{10} P$ 表示。红色圆点代表上调表达基因,绿色圆点代表下调表达基因,灰色圆点代表没有显著变化的基因。

图 6-10　PFOA+0.2 PS、PFOA+2 PS、PFOA+20 PS 差异表达基因火山图

6.5.2 PS-MPs 粒径对 PFOA 诱导的斑马鱼差异表达基因的 GO 富集结果的影响

通过分析 PFOA+0.2 PS 暴露与 PFOA 暴露相比导致的差异表达基因在 GO 数据库富集情况,进一步了解基因功能。PFOA+0.2 PS 暴露组与 PFOA 暴露组相比,分别从生物过程、细胞组分和分子功能三个类别中各选取显著性最

高的十个条目。最显著的生物过程条目分别是造血干细胞迁移（GO：0035701）、肌动蛋白细胞骨架组织（GO：0030036）、基于肌动蛋白丝的过程（GO：0030029）、肌肉结构发育（GO：0061061）、横纹肌细胞发育（GO：0055002）、肌肉细胞分化（GO：0042692）、骨骼肌组织发育（GO：0007519）、肌肉细胞发育（GO：0055001）、骨骼肌器官发育（GO：0060538）和成肌细胞增殖的调控（GO：2000291）；最显著的细胞组分条目为肌节（GO：0030017）、肌原纤维（GO：0030016）、收缩纤维（GO：0043292）、可收缩纤维部分（GO：0044449）、A 组合（GO：0031672）、M 组合（GO：0031430）、肌浆网膜（GO：0033017）、Cul3-RING泛素连接酶复合物（GO：0031463）、肌浆网（GO：0016529）和肌动蛋白细胞骨架（GO：0015629）；最显著的分子功能条目为肌动蛋白结合（GO：0003779）、细胞骨架蛋白结合（GO：0008092）、化学引诱剂活性（GO：0042056）、γ-谷氨酰环转移酶活性（GO：0003839）、趋化因子活性（GO：0008009）、趋化因子受体结合（GO：0042379）、G 蛋白偶联受体结合（GO：0001664）、CXCR 趋化因子受体结合（GO：0045236）、药物跨膜转运蛋白活性（GO：0015238）和转移酶活性，转移氨基酰基（GO：0016755），见表6-6。对差异表达基因急性 GO 二级分类分析[图 6-11(a)]，PFOA＋0.2 PS 暴露组与 PFOA 暴露组相比，斑马鱼的差异表达基因显著富集于细胞过程（上调54，下调133）、单体过程（上调39，下调120）、生物调节（上调33，下调92）、生物过程调控（上调33，下调85）等生物过程；作用于结合（上调63，下调120）、催化活性（上调22，下调42）、核酸结合转录因子活性（上调10，下调14）等分子功能；影响细胞（上调33，下调85）、细胞部分（上调33，下调85）、膜（上调19，下调64）、细胞器（上调27，下调52）等细胞组分。

表6-6　PFOA＋0.2 PS 与 PFOA 相比差异表达基因 GO 富集结果

GO Term	Database	ID	P 值
hematopoietic stem cell migration	生物过程	GO：0035701	7.20×10^{-6}
actin cytoskeleton organization	生物过程	GO：0030036	9.78×10^{-6}
actin filament-based process	生物过程	GO：0030029	1.17×10^{-5}
muscle structure development	生物过程	GO：0061061	2.10×10^{-5}
striated muscle cell development	生物过程	GO：0055002	4.21×10^{-5}
muscle cell differentiation	生物过程	GO：0042692	5.96×10^{-5}
skeletal muscle tissue development	生物过程	GO：0007519	6.00×10^{-5}
muscle cell development	生物过程	GO：0055001	6.84×10^{-5}

表 6-6(续)

GO Term	Database	ID	P 值
skeletal muscle organ development	生物过程	GO:0060538	9.76×10^{-5}
regulation of myoblast proliferation	生物过程	GO:2000291	0.000 150 07
sarcomere	细胞组分	GO:0030017	1.91×10^{-11}
myofibril	细胞组分	GO:0030016	2.54×10^{-11}
contractile fiber	细胞组分	GO:0043292	3.35×10^{-11}
contractile fiber part	细胞组分	GO:0044449	3.35×10^{-11}
A band	细胞组分	GO:0031672	7.40×10^{-7}
M band	细胞组分	GO:0031430	1.82×10^{-6}
sarcoplasmic reticulum membrane	细胞组分	GO:0033017	3.53×10^{-5}
Cul3-RING ubiquitin ligase complex	细胞组分	GO:0031463	6.25×10^{-5}
sarcoplasmic reticulum	细胞组分	GO:0016529	0.000 144 444
actin cytoskeleton	细胞组分	GO:0015629	0.000 150 889
actin binding	分子功能	GO:0003779	3.01×10^{-5}
cytoskeletal protein binding	分子功能	GO:0008092	0.000 330 575
chemoattractant activity	分子功能	GO:0042056	0.000 446 581
gamma-glutamylcyclotransferase activity	分子功能	GO:0003839	0.002 179 419
chemokine activity	分子功能	GO:0008009	0.003 426 24
chemokine receptor binding	分子功能	GO:0042379	0.003 426 24
G-protein coupled receptor binding	分子功能	GO:0001664	0.003 687 699
CXCR chemokine receptor binding	分子功能	GO:0045236	0.004 003 176
drug transmembrane transporter activity	分子功能	GO:0015238	0.006 330 964
transferase activity, transferring amino-acyl groups	分子功能	GO:0016755	0.009 137 458

通过分析 PFOA+2 PS 暴露与 PFOA 暴露相比导致的差异表达基因在 GO 数据库富集情况,进一步了解基因功能。PFOA+2 PS 暴露组与 PFOA 暴露组相比,分别从生物过程、细胞组分和分子功能三个类别中各选取显著性最高的十个条目。富集最显著的生物过程条目分别是跨膜转运(GO:0055085)、单体运输(GO:0044765)、离子运输(GO:0006811)、单体定位(GO:1902578)、酰胺转运(GO:0042886)、阴离子运输(GO:0006820)、离子跨膜转运(GO:0034220)、肽转运(GO:0015833)、一元羧酸运输(GO:0015718)和单价无机阳离子稳态(GO:0055067);富集最显著的细胞组分条目为质膜固有成分(GO:

第6章 聚苯乙烯微塑料对全氟辛酸斑马鱼毒性影响机制

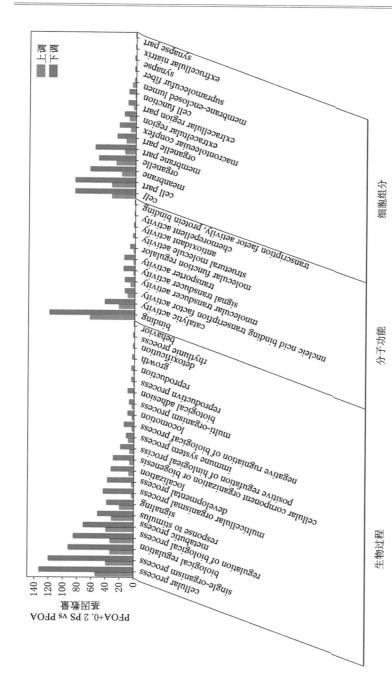

(a) PFOA+0.2 PS

图6-11 PFOA+0.2 PS、PFOA+2 PS和PFOA+20 PS暴露下差异表达基因GO二级分类统计图

横坐标根据生物过程、分子功能和细胞组分进行分类，纵坐标为该条目富集的基因数；红色部分为上调基因，绿色部分为下调基因

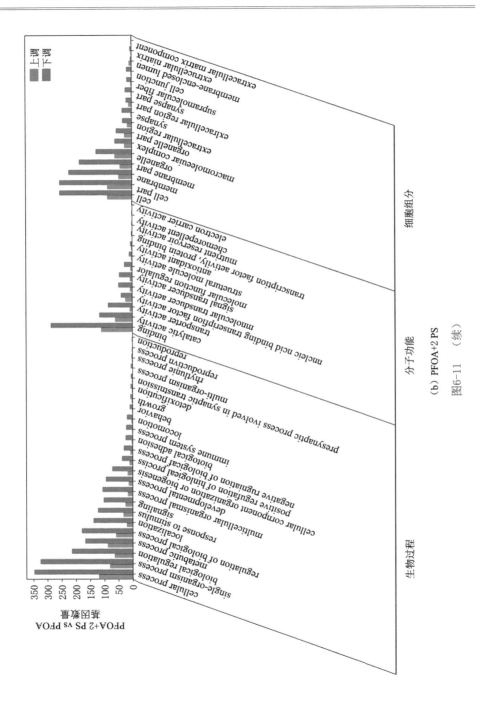

(b) PFOA+2 PS

图6-11 (续)

第6章 聚苯乙烯微塑料对全氟辛酸斑马鱼毒性影响机制

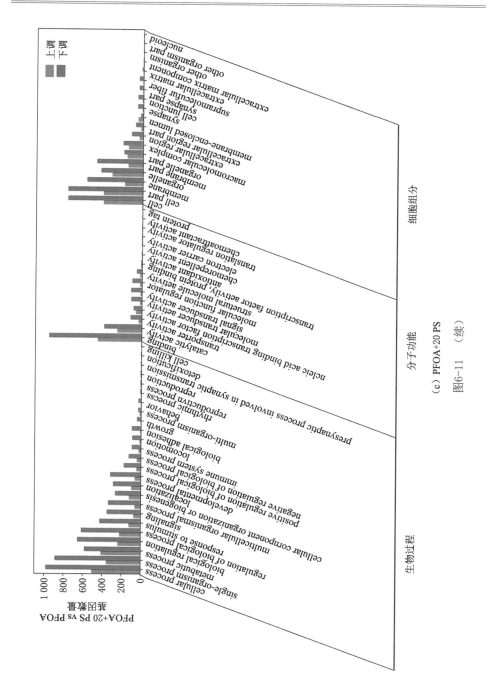

(c) PFOA+20 PS

图6-11 （续）

0031226)、质膜的组成部分(GO:0005887)、细胞外围(GO:0071944)、质膜(GO:0005886)、质膜部分(GO:0044459)、血红蛋白复合体(GO:0005833)、膜的组成部分(GO:0016021)、膜的固有成分(GO:0031224)、顶端质膜(GO:0016324)和细胞顶端(GO:0045177);富集最显著的分子功能条目为活性跨膜转运蛋白活性(GO:0022804)、次级活性跨膜转运蛋白活性(GO:0015291)、跨膜转运蛋白活性(GO:0022857)、底物特异性跨膜转运蛋白活性(GO:0022891)、底物特异性转运蛋白活性(GO:0022892)、离子跨膜转运蛋白活性(GO:0015075)、转运活性(GO:0005215)、阴离子跨膜转运蛋白活性(GO:0008509)、同转运体活性(GO:0015293)和阳离子跨膜转运蛋白活性(GO:0008324),见表6-7。对差异表达基因急性GO二级分类分析[图6-11(b)],PFOA+2 PS暴露组与PFOA暴露组相比,斑马鱼的差异表达基因显著富集于细胞过程(上调119,下调347)、单体过程(上调80,下调325)、生物调节(上调63,下调214)、代谢过程(上调88,下调167)等生物过程;作用于结合(上调109,下调286)、催化活性(上调59,下调115)、转运活性(上调8,下调85)等分子功能;影响细胞(上调86,下调254)、细胞部分(上调86,下调254)、膜(上调46,下调222)、膜部分(上调41,下调185)等细胞组分。

表6-7 PFOA+2 PS 与 PFOA 相比差异表达基因 GO 富集结果

GO Term	Database	ID	P 值
transmembrane transport	生物过程	GO:0055085	1.94×10^{-7}
single-organism transport	生物过程	GO:0044765	8.62×10^{-7}
ion transport	生物过程	GO:0006811	1.08×10^{-6}
single-organism localization	生物过程	GO:1902578	3.08×10^{-6}
amide transport	生物过程	GO:0042886	4.41×10^{-6}
anion transport	生物过程	GO:0006820	1.50×10^{-5}
ion transmembrane transport	生物过程	GO:0034220	1.51×10^{-5}
peptide transport	生物过程	GO:0015833	2.31×10^{-5}
monocarboxylic acid transport	生物过程	GO:0015718	2.37×10^{-5}
monovalent inorganic cation homeostasis	生物过程	GO:0055067	2.39×10^{-5}
intrinsic component of plasma membrane	细胞组分	GO:0031226	5.31×10^{-7}
integral component of plasma membrane	细胞组分	GO:0005887	9.25×10^{-7}
cell periphery	细胞组分	GO:0071944	2.11×10^{-7}
plasma membrane	细胞组分	GO:0005886	4.10×10^{-6}

第6章 聚苯乙烯微塑料对全氟辛酸斑马鱼毒性影响机制

表 6-7(续)

GO Term	Database	ID	P 值
plasma membrane part	细胞组分	GO:0044459	9.35×10^{-6}
hemoglobin complex	细胞组分	GO:0005833	0.001 082
integral component of membrane	细胞组分	GO:0016021	0.002 361
intrinsic component of membrane	细胞组分	GO:0031224	0.002 898
apical plasma membrane	细胞组分	GO:0016324	0.003 521
apical part of cell	细胞组分	GO:0045177	0.005 051
active transmembrane transporter activity	分子功能	GO:0022804	5.80×10^{-12}
secondary active transmembrane transporter activity	分子功能	GO:0015291	2.10×10^{-10}
transmembrane transporter activity	分子功能	GO:0022857	6.66×10^{-10}
substrate-specific transmembrane transporter activity	分子功能	GO:0022891	6.71×10^{-10}
substrate-specific transporter activity	分子功能	GO:0022892	8.68×10^{-10}
ion transmembrane transporter activity	分子功能	GO:0015075	2.46×10^{-9}
transporter activity	分子功能	GO:0005215	3.27×10^{-9}
anion transmembrane transporter activity	分子功能	GO:0008509	1.48×10^{-6}
symporter activity	分子功能	GO:0015293	2.22×10^{-6}
cation transmembrane transporter activity	分子功能	GO:0008324	3.54×10^{-6}

通过分析 PFOA+20 PS 暴露与 PFOA 暴露相比导致的差异表达基因在 GO 数据库富集情况,进一步了解基因功能。PFOA+20 PS 暴露组与 PFOA 暴露组相比,分别从生物过程、细胞组分和分子功能三个类别中各选取显著性最高的十个条目。富集最显著的生物过程条目分别是细胞运动(GO:0048870)、细胞定位(GO:0051674)、移动(GO:0040011)、细胞迁移(GO:0016477)、趋化性(GO:0006935)、巨噬细胞趋化性(GO:0048246)、趋性(GO:0042330)、趋化性调节(GO:0050920)、白细胞趋化性(GO:0030595)和细胞趋化性(GO:0060326);富集最显著的细胞组分条目为中间纤维(GO:0005882)、中间纤维细胞骨架(GO:0045111)、膜的外源成分(GO:0019898)、桥粒(GO:0030057)、核糖体(GO:0005840)、胞质部分(GO:0044445)、质膜的外源成分(GO:0019897)、核糖体亚单位(GO:0044391)、细胞外围(GO:0071944)和 A 组会(GO:0031672);富集最显著的分子功能条目为肌动蛋白结合(GO:0003779)、细胞骨架蛋白结合(GO:0008092)、钙离子结合(GO:0005509)、β 淀粉样蛋白结合(GO:0001540)、结构分子活性(GO:0005198)、酰胺跨膜转运蛋白活性(GO:0042887)、脂肪酶抑制剂活性(GO:0055102)、磷脂酶抑制剂活性(GO:0004859)、

金属离子结合(GO:0046872)和阳离子结合(GO:0043169),见表6-8。对差异表达基因在GO二级分类情况进行分析[图6-11(c)],PFOA+20 PS暴露组与PFOA暴露组相比,斑马鱼的差异表达基因显著富集于细胞过程(上调504,下调979)、单体过程(上调350,下调883)、代谢过程(上调406,下调581)、生物调节(上调233,下调652)等生物过程;作用于结合(上调450,下调955)、催化活性(上调251,下调386)、转运活性(上调31,下调115)等分子功能;影响细胞(上调399,下调769)、细胞部分(上调399,下调769)、膜(上调183,下调575)、细胞器(上调312,下调428)等细胞组分。

表6-8 PFOA+20 PS与PFOA相比差异表达基因GO富集结果

GO Term	Database	ID	P 值
cell motility	生物过程	GO:0048870	8.02×10^{-5}
localization of cell	生物过程	GO:0051674	8.02×10^{-5}
locomotion	生物过程	GO:0040011	8.85×10^{-5}
cell migration	生物过程	GO:0016477	9.07×10^{-5}
chemotaxis	生物过程	GO:0006935	0.000 104
macrophage chemotaxis	生物过程	GO:0048246	0.000 184
taxis	生物过程	GO:0042330	0.000 23
regulation of chemotaxis	生物过程	GO:0050920	0.000 301
leukocyte chemotaxis	生物过程	GO:0030595	0.000 511
cell chemotaxis	生物过程	GO:0060326	0.000 677
intermediate filament	细胞组分	GO:0005882	0.000 242
intermediate filament cytoskeleton	细胞组分	GO:0045111	0.000 242
extrinsic component of membrane	细胞组分	GO:0019898	0.000 96
desmosome	细胞组分	GO:0030057	0.001 58
ribosome	细胞组分	GO:0005840	0.002 112
cytosolic part	细胞组分	GO:0044445	0.002 403
extrinsic component of plasma membrane	细胞组分	GO:0019897	0.003 436
ribosomal subunit	细胞组分	GO:0044391	0.003 905
cell periphery	细胞组分	GO:0071944	0.004 235
A band	细胞组分	GO:0031672	0.004 804
actin binding	分子功能	GO:0003779	3.56×10^{-6}
cytoskeletal protein binding	分子功能	GO:0008092	0.000 303

表 6-8(续)

GO Term	Database	ID	P 值
calcium ion binding	分子功能	GO:0005509	0.000 365
beta-amyloid binding	分子功能	GO:0001540	0.000 882
structural molecule activity	分子功能	GO:0005198	0.001 493
amide transmembrane transporter activity	分子功能	GO:0042887	0.002 268
lipase inhibitor activity	分子功能	GO:0055102	0.002 268
phospholipase inhibitor activity	分子功能	GO:0004859	0.002 268
metal ion binding	分子功能	GO:0046872	0.002 358
cation binding	分子功能	GO:0043169	0.003 096

6.5.3 PS-MPs 粒径对 PFOA 诱导的斑马鱼差异表达基因的通路富集结果的影响

通过分析 PFOA+0.2 PS 与 PFOA 相比时差异表达基因在 KEGG、Reactome 和 Panther 数据库中参与功能通路的情况,进一步了解差异表达基因与生物学通路的关联[图 6-12(a)]。基于显著性筛选富集最显著的前 15 个通路($P<0.05$)。与 PFOA 相比,PFOA+0.2 PS 暴露时最显著的前 15 个通路是趋化因子受体结合趋化因子(R-DRE-380108)、烟碱乙酰胆碱受体信号通路(P00044)、嘌呤再利用(R-DRE-74217)、肽配体结合受体(R-DRE-375276)、G alpha(i)信号事件(R-DRE-418594)、A/1 类(红细胞色素样受体)(R-DRE-373076)、脯氨酸生物合成(P02768)、溶酶体寡糖分解代谢(R-DRE-8853383)、肽链延伸(R-DRE-156902)、Rho GTPase 对细胞骨架的调节(P00016)、嘌呤代谢(R-DRE-73847)、心肌收缩(DRE04260)、AKT 介导的 FOXO1A 失活(R-DRE-211163)、松弛素受体(R-DRE-444821)和 β 细胞基因表达的调控(R-DRE-210745)。

通过分析 PFOA+2 PS 与 PFOA 相比时差异表达基因在 KEGG、Reactome 和 Panther 数据库中参与功能通路的情况,进一步了解差异表达基因与生物学通路的关联[图 6-12(b)]。与 PFOA 相比,PFOA+2 PS 暴露时最显著的前 15 个通路是无机阳离子/阴离子和氨基酸/寡肽的转运(R-DRE-425393)、SLC 介导的跨膜转运(R-DRE-425407)、小分子的跨膜转运(R-DRE-382551)、碳酸氢盐转运蛋白(R-DRE-425381)、红细胞吸收氧气并释放二氧化碳(R-DRE-1247673)、类固醇激素生物合成(DRE00140)、@@红细胞吸收二氧化碳并释放氧气(R-DRE-1237044)、红细胞中 O_2/CO_2 的交换(R-DRE-1480926)、谷胱甘肽代谢(DRE00480)、表面活性剂代谢(R-DRE-5683826)、心肌收缩(DRE04260)、有机

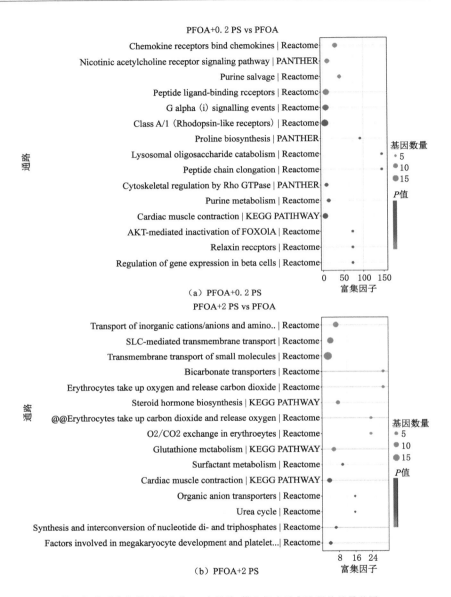

纵坐标表示富集最显著的前 15 个通路,横坐标表示有注释的差异基因占总差异基因的频率与注释的背景基因占总背景基因的频率的比值;颜色代表显著性,颜色越红表示越显著;气泡大小代表富集基因的数量,气泡越大表示富集基因越多。

图 6-12　PFOA+0.2 PS、PFOA+2 PS 和 PFOA+20 PS 暴露时差异表达基因通路富集气泡图

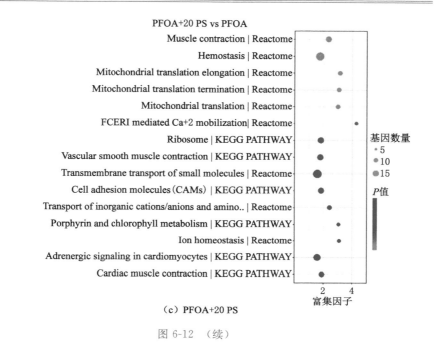

(c) PFOA+20 PS

图 6-12 （续）

阴离子转运体(R-DRE-428643)、尿素循环(R-DRE-70635)、二磷酸核苷酸和三磷酸核苷酸的合成及其相互转化(R-DRE-499943)和巨核细胞发育和血小板生成的相关因素(R-DRE-983231)。

通过分析 PFOA＋20 PS 与 PFOA 相比时差异表达基因在 KEGG、Reactome 和 Panther 数据库中参与功能通路的情况，进一步了解差异表达基因与生物学通路的关联[图 6-12(c)]。与 PFOA 相比，PFOA＋20 PS 暴露时最显著的前 15 个通路是肌肉收缩(R-DRE-397014)、止血(R-DRE-109582)、线粒体翻译延伸(R-DRE-5389840)、线粒体翻译终止(R-DRE-5419276)、线粒体翻译(R-DRE-5368287)、FCERI 介导的 Ca^{2+} 动员(R-DRE-2871809)、核糖体(DRE03010)、血管平滑肌收缩(DRE04270)、小分子的跨膜转运(R-DRE-382551)、细胞黏附分子(DRE04514)、无机阳离子/阴离子和氨基酸/寡肽的转运(R-DRE-425393)、卟啉与叶绿素代谢(DRE00860)、离子稳态(R-DRE-5578775)、心肌细胞的肾上腺素能信号传导(DRE04261)和心肌收缩(DRE04260)。

6.5.4 心肌收缩通路分析

通路结果表明，与 PFOA 相比，PFOA＋0.2 PS、PFOA＋2 PS 和 PFOA＋20 PS 均显著影响心肌收缩通路(图 6-13)。

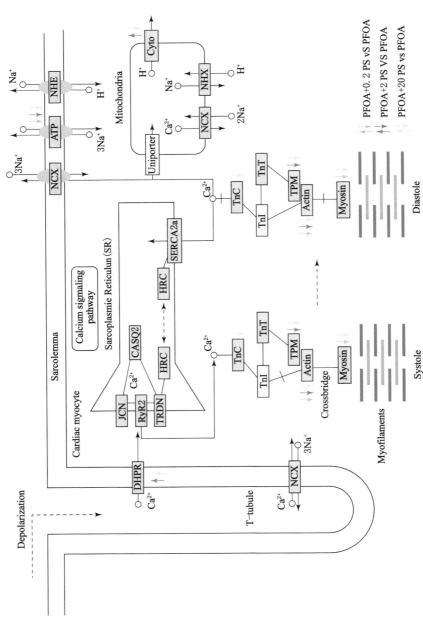

图6-13 心肌收缩通路

蓝色、红色和黄色箭头分别表示与PFOA暴露组相比,斑马鱼在PFOA+0.2 PS、PFOA+2 PS和PFOA+20 PS暴露时的差异表达基因;向上的箭头表示基因上调,向下的箭头表示基因下调。

心脏收缩的力量会产生一种能量波,这种能量波可以测量为血压[15]。2 PS 和 20 PS 影响 Ca^{2+} 运输,2 PS 和 20 PS 导致基因 *cacna1c* 和 *cacng4b* 显著下调,20 PS 导致基因 *cacna2d2b* 显著下调且导致基因 *cacna2d3* 显著上调。2 PS 和 20 PS 显著影响 H^+ 从线粒体向心肌细胞移动,导致基因 *uqcrb* 上调。三种暴露条件下导致收缩期和舒张期肌丝肌动蛋白的基因 *actc1a* 显著下调,同时导致 ATP 转运的基因 *fxyd2* 显著下调。20 PS 导致 ATP 转运的基因 *atp1b3b* 和 *atp1a1b* 显著下调,0.2 PS 和 20 PS 导致 ATP 转运的基因 *atp1a3a* 显著下调。2 PS 和 20 PS 导致基因 *tpm1* 和 *tpm4a* 显著下调,20 PS 导致基因 *tpm4b*、*tnnc1a* 显著下调。对于肌球蛋白,0.2 PS 导致基因 *myh7* 显著下调,20 PS 导致基因 *myl2b* 显著下调。简言之,三种微塑料均影响 ATP 的转运和肌丝的收缩与舒张,2 PS 和 20 PS 还对 Ca^{2+} 跨细胞膜的运输及 H^+ 的跨线粒体运输造成影响。

6.6 本章小结

为研究微塑料对 PFOA 的斑马鱼毒性影响,通过转录组测序技术对 PS-MPs 存在时 PFOA 对斑马鱼毒性机制的影响以及不同粒径 PS-MPs 对斑马鱼毒性影响机制的差异进行研究,得出了以下的结论:

① PFOA 暴露后筛选出 843 个差异表达基因;GO 富集分析表明,PFOA 显著影响的生物过程、细胞组分和分子功能条目多为运输过程、膜组织和转运蛋白活性,这可能与 PFOA 对神经系统的影响有关;通路富集结果表明,PFOA 显著影响神经系统中多种神经递质释放周期通路,包括去甲肾上腺素神经递质释放周期、谷氨酸神经递质释放周期和乙酰胆碱神经递质释放周期通路。

② PS-MPs 存在时的差异表达基因分析表明,PFOA+0.2 PS、PFOA+2 PS 和 PFOA+2 PS 暴露后诱导的差异表达基因分别为 222 个、500 个和 370 个;GO 富集分析结果类似,对细胞过程、单体过程、生物调节等生物过程,结合、催化活性、转运活性等分子功能,膜和细胞等细胞组分造成显著影响;通路富集结果表明,PFOA+0.2 PS 会对神经系统中内整流 K^+ 通道造成显著影响,包括 K^+ 从膜外到膜内的被动运输及 K^+ 从膜内到膜外的主动运输。PFOA+2 PS 和 PFOA+20 PS 显著影响氨基酸代谢,对影响甘氨酸、丝氨酸和苏氨酸代谢通路均造成显著影响,包括丝氨酸的激活及甘氨酸与丝氨酸间的转化。

③ 以 PFOA 暴露为对照组的差异表达基因分析表明,PFOA+0.2 PS 暴露诱导了 390 个差异表达基因,PFOA+2 PS 暴露诱导了 869 个差异表达基因,PFOA+20 PS 暴露诱导了 2 854 个差异表达基因;而 GO 富集分析表明,

PS-MPs显著影响基因显著富集于细胞过程、单体过程、生物调节等生物过程,结合、催化活性等分子功能,细胞、膜和细胞器等细胞组分;通路富集结果则表明,三种粒径的 PS-MPs 均影响 ATP 的转运和肌丝的收缩与舒张,2 PS 和 20 PS 还对 Ca^{2+} 跨细胞膜的运输及 H^+ 的跨线粒体运输造成影响。

参考文献

[1] ZHANG X W, XIA P, WANG P P, et al. Omics advances in ecotoxicology [J]. Environmental science and technology, 2018, 52(7): 3842-3851.

[2] ZIV N E, GARNER C C. Presynaptic development and active zones[M]// Encyclopedia of Neuroscience. Amsterdam: Elsevier, 2009: 957-966.

[3] HARRIS K P, LITTLETON J T. Transmission, development, and plasticity of synapses[J]. Genetics, 2015, 201(2): 345-375.

[4] DENKER A, BETHANI I, KRÖHNERT K, et al. A small pool of vesicles maintains synaptic activity *in vivo*[J]. Proceedings of the national academy of sciences of the United States of America, 2011, 108(41): 17177-17182.

[5] GÓMEZ-CANELA C, TORNERO-CAÑADAS D, PRATS E, et al. Comprehensive characterization of neurochemicals in three zebrafish chemical models of human acute organophosphorus poisoning using liquid chromatography-tandem mass spectrometry[J]. Analytical and bioanalytical chemistry, 2018, 410(6): 1735-1748.

[6] BYLUND D B. Norepinephrine[J]. Encyclopedia of the neurological sciences, 2001(49): 980-1001.

[7] PARK E, VELUMIAN A A, FEHLINGS M G. The role of excitotoxicity in secondary mechanisms of spinal cord injury: a review with an emphasis on the implications for white matter degeneration[J]. Journal of neurotrauma, 2004, 21(6): 754-774.

[8] HAUG M F, GESEMANN M, MUELLER T, et al. Phylogeny and expression divergence of metabotropic glutamate receptor genes in the brain of zebrafish (*Danio rerio*)[J]. Journal of comparative neurology, 2013, 521(7): 1533-1560.

[9] SEIBT K J, OLIVEIRA R D, RICO E P, et al. Typical and atypical antipsychotics alter acetylcholinesterase activity and ACHE expression in zebrafish (*Danio rerio*) brain[J]. Comparative biochemistry and physiology toxicology and pharma-

cology,2009,150(1):10-15.

[10] GOLDIN A L.Neuronal channels and receptors[M]//Molecular Neurology.Amsterdam:Elsevier,2007.

[11] LEE H J,CHOI T I,KIM Y M,et al.Regulation of habenular G-protein gamma 8 on learning and memory via modulation of the central acetylcholine system[J].Molecular psychiatry,2021,26:3737-3750.

[12] TSUJI-TAMURA K,SATO M,FUJITA M,et al.Glycine exerts dose-dependent biphasic effects on vascular development of zebrafish embryos [J].Biochemical and biophysical research communications,2020,527(2):539-544.

[13] HE L Q,DING Y Q,ZHOU X H,et al.Serine signaling governs metabolichomeostasis and health[J].Trends in endocrinology and metabolism,2023,34(6):361-372.

[14] KOHLMEIER M. Threonine[M]//Nutrient Metabolism. Amsterdam:Elsevier,2003.

[15] WATKINS A.24-The electrical heart:energy in cardiac health and disease [J].Energy medicine east and west,2011(1):305-317.